# Reckoning Words

William Marshall's title page to Gilbert Wats's translation of Bacon's *De augmentis*, 1640. Courtesy of The Pierpont Morgan Library, New York. PML 37241.

# Reckoning Words

Baconian Science and the Construction of Truth in English Renaissance Culture

Diana B. Altegoer

Madison • Teaneck
Fairleigh Dickinson University Press
London: Associated University Presses

© 2000 by Associated University Presses, Inc.

All rights reserved. Authorization to photocopy items for internal or personal use, or the internal or personal use of specific clients, is granted by the copyright owner, provided that a base fee of $10.00, plus eight cents per page, per copy is paid directly to the Copyright Clearance Center, 222 Rosewood Drive, Danvers, Massachusetts 01923. [0-8386-3825-2/00 $10.00 + 8¢ pp, pc.]

Associated University Presses
440 Forsgate Drive
Cranbury, NJ 08512

Associated University Presses
16 Barter Street
London WC1A 2AH, England

Associated University Presses
P.O. Box 338, Port Credit
Mississauga, Ontario
Canada L5G 4L8

The paper used in this publication meets the requirements of the American National Standard for Permanence of Paper for Printed Library Materials Z39.48-1984.

**Library of Congress Cataloging-in-Publication Data**

Altegoer, Diana B, 1962–
   Reckoning words : Baconian science and the construction of truth in English Renaissance culture / Diana B. Altegoer.
     p. cm.
   Includes bibliographical references and index.
   ISBN 0-8386-3825-2 (alk. paper)
   1. English literature—Early modern, 1500–1700—History and criticism.  2. Bacon, Francis, 1561–1626—Contributions in philosophy of nature.  3. Literature and science—England—History—17th century.  4. Science—England—History—17th century.  5. Bacon, Francis 1561–1626—Influence.  6. Knowledge, Theory of, in literature.  7. Philosophy of nature in literature.  8. Renaissance—England.  9. Reality in literature.  10. Truth in literature.  I. Title.

PR438.S37 A45 2000
808'.09'031—dc21
                                              99-055050

PRINTED IN THE UNITED STATES OF AMERICA

To my parents,

Helen Altegoer,
and in memory of
Hans W. Altegoer
(1927–1976)

# Contents

| | |
|---|---:|
| Acknowledgments | 9 |
| Bibliographical Note | 11 |
| Prologue | 13 |
| 1. Renaissance *Res* and *Verba:* Toward a Poetics of Truth | 29 |
| 2. *Copia Verborum:* The Maker's Knowledge in Renaissance Poetics and Rhetoric | 51 |
| 3. Francis Bacon: The "Restauration" | 76 |
| 4. Bacon's Politics of Allegory | 95 |
| 5. Wise Men's Counters: Visual and Verbal Knowledge in Hobbes and Boyle | 113 |
| 6. The Figure in the Pool: Milton's Epistemology of Nature | 137 |
| Epilogue | 167 |
| Notes | 173 |
| Bibliography | 191 |
| Index | 205 |

# Acknowledgments

THIS WORK GREW OUT OF THE LECTURES AND TUTORIALS ON Renaissance rhetorical theory given by Dr. Avril Bruten of St. Hugh's College during my first year of doctoral work at Oxford University. My warmest thanks are due to her guidance as dissertation supervisor, for providing continual intellectual nourishment and invaluable "food" for thought. My students in the English departments at Vassar College and Old Dominion University also deserve thanks for their sympathetic encouragement through difficult years of teaching and writing. I would also like to acknowledge and thank the Awards Committee and Bursary of Vassar College, the British Overseas Research Awards Committee for their financial assistance, as well as the Committee for Graduate Studies at Oxford and the Fellows of Merton College for awarding several travel grants. Old Dominion University also awarded me several summer research fellowships. Portions of chapters three and four appeared in *Renaissance Papers 1993*.

Thanks also go to the curators of the Pierpont Morgan Library, and to Despina Coutavas from the Department of Photography and Rights, for their efficiency and service. Christine Retz and the editors at AUP were also extremely helpful. For their friendship, conversation, and timely support, I am indebted to my colleagues in Britain: Madeleine Fontana Barrows, Gary Barrows, Georgia Brown, Adam Smith, Tony Trowles, Maureen McEleron Watry, Paul Watry; and at Vassar: Margaret Fusco, Donna Heiland, and Susan Zlotnick. My deepest thanks are due to my mother, for her constant encouragement, support and conversation that extended beyond the merely rhetorical.

# Bibliographical Note

References to Bacon's texts are taken from *The Philosophical Works of Francis Bacon,* edited by John M. Robertson (New York, 1905), *The Philosophy of Francis Bacon,* edited by Benjamin Farrington (Liverpool, 1970), and *The Advancement of Learning and the New Atlantis,* edited by Arthur Johnston (Oxford, 1974), and finally, *The Essays,* edited by John Pitcher (Harmondsworth, Middlesex, 1985). References to Bacon's correspondence and letters are taken from *The Letters and Life of Francis Bacon,* published as vols. VIII–XIV in *The Works of Francis Bacon,* 14 vols., edited by James Spedding, R. L. Ellis, and D. D. Heath (New York: Garrett Press, 1968 reprint).

References to page numbers of specific works taken from these editions are given in round brackets in the text, for the sake of convenience. I have cited the English translations from the Latin texts, provided by Farrington and Spedding. In citing from the specific works, I have used the following abbreviations: from Robertson's edition, *Novum Organon* (cited as *NO*), *De Dignitate et Augmentis Scientiarum* (*DAS*), and *De Sapientia Veterum* (*DSV*); from Farrington's edition, *The Masculine Birth of Time* (*MBT*), *Thoughts and Conclusions* (*TC*), and *The Refutation of Philosophies* (*RP*); and from Johnston's edition, *The Advancement of Learning* (*ADV*), and the *New Atlantis* (*NA*).

All references to Augustine's texts are taken from the following editions and are cited in round brackets after the quotation. *The Confessions* (cited as *Conf*) translated by R. S. Pine-Coffin (Harmondsworth, Middlesex, 1961), *De Magistro, or Concerning the Teacher* (*DM*) translated by George G. Leckie (London, 1939), *De Doctrina Christiana* (*DDC*) translated by D. W. Robertson, Jr. (New York, 1958).

# Prologue

> [T]he exact study of languages, and the efficacy of preaching, did bring in an affectionate study of eloquence and copie of speech, which then began to flourish. This grew speedily to an excess; for men began to hunt more after words than matter.
> —Francis Bacon, *The Advancement of Learning* (Book I, iv, 2)

FEW CATCHPHRASES HAD SUCH EMOTIONAL APPEAL FOR EARLYmodern writers as this one, *res et verba,* subject matter and verbal construction, which signaled the debate concerning the use of copious eloquence and language to represent philosophical truth. This conflict between rhetoric and philosophy would touch on the religious and psychological pulse of writers as diverse as Francis Bacon, Thomas Hobbes, Robert Boyle, and John Milton. The issue at stake involved the notion of truth and judgment that would later prove central to the resolution of civil war divisiveness and linguistic confusion. Political and epistemological parties were erected and enemies discovered based upon their propensity to figure metaphors into their writing and thinking. Masculine toughness was pitted against a weakened and decidedly feminine taste for eloquence and beauty. The remarkable love of ornate and oral drama would be held in suspicion and contempt. By the end of the seventeenth century, *res et verba* would become conceptually dislocated. *Res* and *verba* (words and things) would be set apart in an epistemological contract that located meaning solely within the referent, the material objects of the natural world.

The following historical reconstruction of Baconian science (as it arose in the seventeenth century) places emphasis on the verbal and political construction of knowledge. Certainly, Bacon helped to inaugurate that conceptual split between the arts of rhetoric and language and the study of natural philosophy, positing that *things* should be separated from *words* in order to fully understand the workings of nature, producing thereby real knowledge. Yet, the continued employment of myth, paradox, and fable in his philosophical writings shows that Bacon was still enmeshed in that rhetorical frame of mind which constantly reassessed philological possibilities in order to discover mental processes and therein the reflected truths of nature. Given this contradiction at the center of Bacon's writing and thinking, the first issue remains: how to define the notion of Baconian science? The traditional view of Bacon holds that the philosopher posits a strict separation between the study of history, philosophy, and poetry, a strict separation between words and things, in order to advance his experimental method premised upon a mechanical understanding of nature, of things and works. Bacon is clear that, to understand anything, one must clearly understand things as opposed to words.

In their *Dialectic of Enlightenment,* Max Horkheimer and Theodor Adorno cite Bacon as the main culprit in the project of the Enlightenment, whose aim it was to dissolve fable and to replace this myth with knowledge.[1] According to Horkheimer and Adorno, Bacon argued that the mind of man and the nature of things are held in disharmony by the vain notions of past masters and present advocates who cannot imagine a world not directly found and mentioned in the books of the past. The invention of the printing press and artillery, and the discovery of the compass in navigation, proved to be the three breakthroughs that succeeded in defining the modern mind, and all three happened by chance. They were not the result of any labored reading of the great authors, Plato, Aristotle, Augustine, or Aquinas. These discoveries were launched through luck and fortune and through the endurance of the virtuoso individual, propelled by the Machiavellian notion of virtu and fortune, luck and daring.

In the late twentieth century, the importance of these three inventions is obvious—in the world of knowledge, learning, war, politics, and international markets. Bacon envisaged knowledge as power; popes, kings, philosophers, and soldiers could not control technology until they were able to buy and pay for those individuals who held mastery over *things* and not merely over the word, the sentence, and the idea.

> Following a tradition that extends back through Saint-Simon to Bacon the decisionistic definition of the relation of expertise to political prac-

tice is being abandoned by many in favor of a *technocratic model*. The dependence of the professional on the politician appears to have reversed itself. The latter becomes the mere agent of a scientific intelligentsia, which, in concrete circumstances, elaborates the objective implications and requirements of available techniques and resources as well as of optimal strategies and rules of control.... The state seems forced to abandon the substance of power in favor of an efficient way of applying available techniques in the framework of strategies that are objectively called for.... It becomes instead the organ of thoroughly rational administration.[2]

Nature (things) would henceforth be mined and explored, the scientist, according to Carolyn Merchant, "searching out her secrets for human improvement."[3]

Paige Dubois, following perhaps the tradition begun by Adorno and Horkheimer and completed by Merchant, argues that "Bacon is making the world safe for torture by presenting a scientific program that assumes violence and coercion to be essential for the domination of nature."[4] According to this view, Bacon replaced myth with philosophy, animism with monotheism, and posited a (masculine) observer who dominated, objectified, and alienated human existence by separating and distancing the witness from nature.[5] However, Merchant also acknowledges that Bacon, notably in the preface to "The Great Instauration," voiced self-doubts about the use of torture, where "the reader can detect a self-reflexive, contradictory impulse in the text" with Bacon noticing the material, the sensual, the pleasurable in nature.[6] Nineteenth-century Marxists such as Engels, who may have admired Bacon's mastery of nature, thought his reasoning too aphoristic, allegorical, poetic (despite his ambivalence to poetry).

My own study, in contrast to that of the cultural materialists listed above, will consist of a rhetorical reading of Bacon's scientific reasoning. Because of the various contradictions in the reception of Bacon, it is worthwhile to explore the complexities of Baconian science, to consider its resources in classical rhetoric, and the cultural and linguistic habits of thought in the early seventeenth century. While not idealizing Bacon's project, or claiming it as liberatory, I contend that Bacon did not call that fissure of science and the arts into being; rather, he conceptualized a unique relationship between the two by creating an experiential (and rhetoricized) "logic" that allowed nature to shape and fashion the perceiving mind of the witness. By calling into question the usual way of dividing disciplines into logic, rhetoric, and poetics, Bacon suggested the constructed nature of these disciplines, and of traditional forms of knowledge. For late-sixteenth-century writers, logic could not readily be separated from rhetoric. Perhaps because of the rhetorical and public nature of this scientific logic, the observer then transmitted this knowledge using verbal representations that were literary in character and political in function. As such, logic was as closely related to eloquence and artful thinking as was rhetoric, in order for the practitioner to achieve the status of a gentleman (and a civic statesman) in Elizabethan and Stuart England.

In many ways, Bacon claimed that logic and rhetoric should not be separated. He was perhaps the first to intuit clearly that truth and knowledge would be served by linking it with (political) power. He attempted to find a philosophical position that avoided the kind of authoritarian practices of the University Schoolmen who continually relied on Aristotelian notions of scientific investigation. In the *Advancement of Learning,* Bacon validated the hieroglyph as a form of writing that appealed to the imagination rather than to strict reason. Likewise, in *De Sapientia Veterum,* he commended the fables of Homer and Hesiod, which rely on instinct and fantasy, as heuristic fictions that precede argument and Aristotelian intellectualism.

Yet, according to Pérez-Ramos the combination of truth (influenced by his interest in logic) and civic usefulness (rhetoric) in Bacon's writings meant that his program was also anti-authoritarian and utilitarian, thus paving the way for Republican Baconianism, millenarian Baconianism, and even the Enlightenment Baconianism of the French Encyclopaedists.[7] Bacon's fluid political alignments, his early endorsement and later renunciation of the condemned Earls of Essex and Somerset, led him to propose a scientific scheme that stressed the detachment and objectivity of natural philosophers working in collaboration, a plan that held the monarch/patron at a benign but discrete distance from the scientific project. This plan would release the scientist from any attachment to the growing political corruption inherent in the courtly culture of both Elizabeth and James I.

Bacon's idea (that there is a truth separate from verbal structures) served not only to sustain a particular political scheme but to protect those individuals seeking political power and position within the state, and who needed a method that would prove unimpeachable and difficult to dislodge and refute. And, in the later part of the seventeenth century, this ideology would be attached to the name of Bacon. The debate triggered by Bacon early in the century, and which centered on the establishment of truth claims—that is, how a statement could be known and verified—ultimately was coopted by the Royal Society to separate its members from worldly and political concerns. This is not to say that there were no ideological wars between different philosophers (dualists versus monists, Cartesians versus Newtonians, and so on). These philosophers, however, could certainly separate themselves from poets and rhetoricians, and hence interfering politicians, and it was this program of separation that would, ironically, carry the name of Francis Bacon.

The relationship between a Baconian philosophy of nature and the foundational practices of the Royal Society has always been problematic. Thomas Sprat, a supporter of the monarchy, denied that there were any Republican Baconians, when the Cromwellian Baconians ("low Baconians") included such influential midcentury thinkers as Jan Comenius, Samuel Hartlib, John Webster, and John Wilkins, all of whom used Bacon in an attempt to distance themselves from Anglican and Royalist clergy, as well as the University Schoolmen. For these nonconformists, the reevaluation of the mechanical arts served as a means of moral self-perfection. Their opponents in the Royal Society, such as Joseph Glanville, claimed Bacon's *New Atlantis* as the model for the Society, arguing (in an un-Baconian way) in favor of a new criterion of moral certainty linked to the degree of reliability of the observing witness. The Society, unlike Bacon, accepted probabilistic and fallibilist accounts of knowledge premised on an *a priori* assumption of the witness's moral character, which itself was based on his standing (birth) in the political community, hence the rise of the gentleman virtuoso. Certainly not derived from Baconian conceptions of human knowledge, these notions succeeded in placing the scientific enterprise within a theological, judicial, and social debate that could not adequately be resolved by the experimental method based on detached observation developed by Bacon and supposedly perfected by Royal Society scientists such as Robert Boyle (Thomas Hobbes would successfully anatomize these contradictions in his public disputes with Boyle).

Bacon's system of knowledge recognized the construction of knowledge based on principles of rhetoric and poetics and aligned with a new kind of scientific demonstration. Although a fear of figurative language animated the philosophical discourse of the period, as well as Bacon's own writings, and explains, in part, the emphasis on collaboration and the detached observer of nature, the body of the witness—the five senses—also was acknowledged to be involved in the observation of nature; hence, Bacon's elaborate system of the idols of the mind to correct any error caused by a flawed and fallen body. In Bacon's work, the dominant and masculine language of truth, reason, detachment (indifference), and observation continued deliberately to negotiate with a discourse of fiction, emotion, intersubjectivity, political networks, passion, and feeling. Only in such a dialectical interchange can the mirror of the mind be polished in order for truth to be pursued, if not ultimately decided.

Antonio Pérez-Ramos' important study of Bacon considers the influence of the maker's knowledge tradition in Bacon's philosophy. This tradition can be traced to classical antiquity; it "postulates an intimate relationship between objects of cognition and objects of construction, and regards knowing as a kind of making or as a capacity to make (*verum factum*)."[8] The motto would be "I know x because I made/did x," and it can be distinguished from a user's knowledge (Plato, in *Cratylus*) or a beholder's knowledge.[9] The prospective knower can impose a certain conceptual frame on things, in that they are potential objects of technical (re)production (*Idea*, 49–50).

Following from this idea, Baconian science has been argued as an enterprise using a utilitarian rhetoric to advocate practical reason and technologically exploitable knowledge in natural inquiries. With this kind of technocratic view, Pérez-Ramos argues,

> rationality is made coextensive with practical and technological reason, as it appears embodied in the natural sciences, and no longer stands in need of further sanction. A self-contained conception of human reason, if exclusively exhibited in technological assets, hardly needs a sanction and its very success discourages criticism (*Idea*, 4).

However, this worldview, while adequate in describing the practices of the Royal Society, was perhaps not envisioned or expressed by Bacon himself, who could not have foreseen how power derived from knowledge would lead to the complete domination of nature. Instead, Bacon envisioned a program constituted in the autonomous character of practical reason, in which the ideals of contemplation (pure science and moral philosophy) were balanced by ideals of action (applied science and praxis). Or, as Bacon writes in Aphorism 124 of the *Novum Organon:* "truth and utility are the very things themselves (*ipsissimae res*); so works themselves are of greater value as pledges of truth than as comforts of life."

In the following work, I will first be looking at the various patterns of thought that existed and operated in the early seventeenth century to determine the inception of Baconian ideals of truth and praxis (usefulness). Second, I will try to identify the diverging genealogical positions that supposedly were prompted by the various meanings of Baconianism in the latter part of the century. In formulating these ideals of knowledge, Bacon was influenced not by only the maker's knowledge tradition but by humanist views on logic (in the pursuit of natural philosophy) and rhetoric as expressed by contemporary poets, painters, sculptors, and architects. Although Bacon declares *verba* (words) to be secondary to *res* (things), he continues to rely on a rhetorical (and poetical) understanding of constructing verbal and physical objects of nature. The primary goal of this work, then, is to recover the meaning of *res et verba* (subject matter and expression on language) in the rhetoric of several sixteenth- and seventeenth-century writers, using as a pivotal point the views on the discovery and transmission of truth expressed by Francis Bacon.

Bacon proved to be a transitional figure in this debate regarding the representation of things through language, and the various claims made by rhetoricians and philosophers. In employing the hermeneutics of both an Augustinian and an Erasmian sign system, Bacon laid the foundation of two distinct approaches to the creation of meaning in language. One approach adopted the terms set by Plato in the *Republic,* whereby words were designed and used as mere copies of things. The second approach saw language as a constructed performance rather than as a copy of reality, an epistemological stance derived ultimately from the idea of dramatized *mimesis* expressed by the pre-Socratic poets and dramatists.

In this second linguistic mode, language creates an emotional and intellectual response in both the reader and the audience. Meaning is not fixed through an arbitrary decision of the reason and the intellect, but is determined by an "affective" experience corresponding in nature to the "affective stylistics" articulated by Stanley Fish and the rhetorical ideal of life argued by Richard Lanham.[10] This dichotomy is often seen as an encounter between two very different modes of life, the rhetorical and the philosophical. *Homo rhetoricus,* the social self, is pitted against *homo seriosus,* the central self. According to Lanham and Fish, this bifurcation of learning necessarily divides knowledge into the playful and the serious, the fictional and the true, the counterfeit and the original.

Barbara Shapiro continually argues, however, that for sixteenth- and seventeenth-century writers, these thickened boundaries, between disciplines and even between epistemological stances, were not always so apparent and simplistic.[11] An awareness of the particular historical situation, she says, suggests that the rhetorical and the philosophical were not always in confrontation. Similarly, my own goal is to blur the boundaries between these strict partitions by determining how a group of writers straddled the epistemological divide by merging rhetoric and linguistic play into a serious philosophical enterprise.

In the first chapter, I set out to determine the various associative ideas that belong to any discussion of *res et verba,* defining the phrase more precisely as it relates to sixteenth- and seventeenth-century language theory and science studies. I set the discussion within the framework of ancient and modern ideas concerning the rule of metaphor and imitation theory. The distinction of *res* and *verba* traditionally suggests the related issues of *mimesis,* and the problems presented by an accurate correspondence (resemblance) between words and things. The center of the debate—whether among the ancients of the Platonic academy, the moderns of the Royal Society, or the postmoderns of the twentieth-century literary and scientific communities—inevitably concerns the preference of a plain style over a more dramatic, copious, and indeterminate style of writing.

The medievals were still influenced by an Aristotelian philosophy of nature that combined with an Augustinean hermeneutical program of reading and interpretation. The sign/signified dichotomy articulated by Augustine in *De Magistro* and *De Doctrina Christiana* resolved the conflict between words and things simply by denying that this dichotomy served an essential purpose in the discovery of truth. Instead, words and things, united in the concept of *res et verba,* combined in a common and useful discursive enterprise which, in the temporality of the human world, matched that intuitive spark Augustine called grace. The ephemeral, fleeting gesture (true *res* or reality) subordinated the more sensuous, and thereby more plodding and hindering, words and things that preceded truth. This transcendent truth was identical to Augustine's Christian God, recognizable through the rituals of the Christian Church. In this sense, final authority was given by fiat; argumentation through dialectic became an exercise in revealing a truth that had already been discovered in the doctrines of the Church.

By the sixteenth century, a peculiar shift in sensibility is apparent. A group of educators and rhetoricians, including Erasmus, changed the terms of Augustine's policy of *res et verba*. The emphasis shifted so that the ephemeral and magical nature inherent in verbal orations came to match, and even stand for, the transcendent and pure *res* of Augustine's doctrine. By elevating the Word, these rhetoricians (or humanists) cast off things from their complementary relationship with words. Human eloquence would take the place of any transcendent truth. However, as with Augustine, the issue of Christ as *logos,* as *verba*—and even, with Erasmus, as *sermo*—neatly resolved any conflict for the sixteenth-century rhetoricians. Truth might be discovered with the efficiency of an oral debate, resembling in nature the *viva voce* examination of university undergraduates. There is no surprise that, in this context, words undermined the efficacy of things. Fantastic statement, detached from any natural phenomena, could be uttered and believed without the restraining influence of either the Roman Church or the natural world. The guarantee rested with the virtuoso ability of the poet-philosopher to skillfully construct a nature that permitted his own advancement in the political and courtly world.

The eclecticism of Renaissance humanism effectively reduced the distinction between poetics, rhetoric, and a dialectic that bound the verbal sciences to the theoretical and practical sciences of philosophy and history. Logic as a scientifically demonstrative discipline gave way to a dialectic indistinguishable from rhetoric. Embellishment and verbal figures were employed for a specific political purpose, not as decoration or illustration, but as necessary elements of proof and argumentation in both disciplines. Rhetoricians such as Puttenham and Wilson used their texts to promote their own interests in the courts of Elizabeth and James. By the seventeenth century, the dichotomies established by ancient Greek and Roman philosophers, the distinction between scientific demonstration and probability, had finally eroded, establishing a mental habit that organized knowledge by investigating probabilities and conjectures, all of which were contrived to legitimize the monarchy and the values of a national and courtly life. It is not surprising, then, that problems of representation and *mimesis,* traditionally limited to poetics, should be central to the concerns of rhetoricians and philosophers of the period.

In an effort to combat the confusion of rhetoric and logic, Francis Bacon attempted in his *Instauratio Magna* a restauration of knowledge founded on new methods of discovery.[12] Bacon, however, did not succeed in fully removing himself from a rhetorical frame of mind, and his attempt to rationalize the relationship between words and things employed the same types of rhetorical and psychological practices found in sixteenth-century rhetorical handbooks. In the *De Dignitate & Augmentis Scientiarum,* Bacon claimed that all human learning flows from the three fountains of memory, imagination, and reason, from which emanate history, poesy, and philosophy; there can be no others. Poesy, aligned with the imagination, held a pivotal place in Bacon's scheme to advance learning; by linking reason with the will and appetite, the poetic imagination creates a pleasurable imitation of history and serves to interpret the hidden wisdom of ancient fables. In grasping the basic principles of imitation, the scientist was able to unveil the figure and reveal the transcendent truth hidden beneath. Bacon thus reappropriated the semantics urged by Erasmus, and added an allegorical and philosophical dimension. For Bacon, the unmasking of myth, and figurative language generally, enabled the yoking of ancient and intuitive wisdom with the advancement of modern learning. Bacon wished to understand this age of fantasy that preceded the age of reason, where fables, a type of hieroglyphic, served as argumentation. Basing his experiential method on a similar, instinctive comprehension of truth, Bacon intended his *Instauratio* to redeem knowledge and restore learning to its primitive (prelapsarian) purity.

In this scheme, metaphor, schemes, tropes, including myth and allegory, were employed to keep truth from becoming "vulgar," to challenge the receiver into strenuous efforts of discovery, and to render truth, once discovered, the more dearly held for the effort it took to discover it. In the *De Sapientia Veterum* (*The Wisdom of the Ancients*), he commended the fables of Homer and Hesiod, which relied on instinct and fantasy, as heuristic fictions that preceded argument and

Aristotelian intellectualism. This mimetic element in philosophical poesis served as a "pleasing clarification," a "total transparency" which moved the audience to a particular mode of action.[13] In his great "instauration," Bacon endorsed two types of logic; one type tried to capture the nature of *things* themselves, while the other attempted to polish the mirror of the mind by stimulating the senses to further inquiry and interpretation. This last type of "logic" relied on rhetorical strategies asking the imagination, in conjunction with reason, to reduce intellectual conceits into sensible images and to re-experience and rewrite fables in light of a new scientific and political pattern.

This intellectualizing of myth proved to be Bacon's greatest influence on subsequent philosophers of nature, language theorists, and poets, albeit without his complexities regarding the symbiotic relationship between scientific understanding and affective poetics. In the manner of Bacon, Restoration writers openly encouraged a plain style based on rational principles and objective knowledge perceptible to the senses. Thomas Sprat separated the knowledge of nature from the colors of rhetoric, the devices of fancy, and the delightful deceit of fables. He further relegated the imagination to the realm of poetry, which was governed not by verity but by verisimilitude. This agenda, which was certainly not practiced with any degree of consistency, served as an epistemological foundation for a broader political scheme. The appeal to reason and plain style was itself a rhetorical strategy used by Royal Society scientists and defenders of the restored Church of England to distinguish themselves from "enthusiastic" nonconformists who abused the rational basis of language.

Even within this group of writers, however, there were discontinuities of thought regarding words and things. In this late period, there were still defenders of the essential status of figures who assumed a close relationship between thought and language. In fact, the Baconian plain style, based on things, continued to be rich in metaphor and analogy. Robert Boyle expressed the need for "apposite comparison" or analogy in his scientific writing, habitually presenting his findings in the form of dialogue and storytelling. John Webster, an avowed Baconian and occultist, wished to penetrate the language of nature to discover the paradisial language of Adam.

By the middle of the seventeenth century, there was a continued preoccupation and anxiety about the usefulness of visual perception in the quest for truth, and the dangerous illusions to be found in any mirrored reflection or copy of reality. Given the separation between science and the arts, a relationship Bacon had been careful to maintain, thinkers at midcentury expressed a growing antithesis between the visual and the verbal faculties, the former privileged by the natural philosopher, the latter by historians, poets, and politicians. In relation to this, the Royal Society, led by Robert Boyle, stressed that the witness to truth must be a particular kind of person and exhibit certain kinds of behavior, such as rationality, detachment, and civic humanism. In fact, the study of nature should reinforce such tendencies in any practitioner, as though nature itself (herself) has a hand in shaping the moral character of the observer and witness. In this, Boyle did indeed follow the thinking of Bacon, without conceptualizing exactly how this shaping could occur. Bacon, in contrast, had used traditional ideas of imitation and the making of forms to theorize how the observation and construction of things might serve as a pedagogical discipline, thereby including the poet, artist and craftsman in his advancement of learning.

By midcentury, however, the renewed emphasis on reason and intellectualism, which Bacon had attacked in the Schoolmen, can be seen in the work of Milton's *Paradise Lost,* whose character of Eve is warned against any flight of idle fancy, and the dangers of gazing with desire into any mirrored image. She must continue to rely on the interpretive and rational skill of her husband Adam in order not to let her insufficient bodily reactions pollute or influence the essential truths of the natural world. In Milton's poem, Raphael's intuitive reason would make the angel an excellent example of the scientific and Christian virtuoso praised by Sprat and Boyle. For Milton, resemblances between heavenly objects, perceived by the angels, are unique and trustworthy; in contrast, likenesses between the fallen objects found in nature, and detected by the corrupted human perception, must be held in suspicion. There can be little play between sign and signified, word and thing, which leads to further inquiry and discovery. The narrow "ideolect" that Milton invents for himself, for his reader, and for general humanity figured in Eve and Adam, is based on a biblical language that is available to any reader chosen and seasoned by God. The intuitive faculty, related to right reason, ensures that this linguistic code is interpreted according to God's just commands. No further inquiry is necessary, for the narrowness of the "ideolect" leads inevitably to truth.

Yet, there are several contradictions and subversions in *Paradise Lost;* I would argue that Eve is a type of empirical scientist who uses rhetorical strategies to uncover the truths of nature. Eve proves herself to be a necessary counterpart to the rational Adam, providing practical knowledge along with the theoretical. Only when the two become separated does the purely rhetorical Satan manage to precipitate the fall of nature and knowledge. Redemption is possible, though, with the reconciliation of logic and rhetoric, theory and practice, thereby making Adam and Eve coequal in the advancement of learning, and partially recovering Bacon's intent to harmonize truth and usefulness. In his use of "affective stylistics," Milton, like Bacon, blurs the boundary between words and things, rhetoric and philosophy; more important was the perceptive eye and mind that were trained constantly to reread, reassess, and, finally, rewrite the ambiguous resemblances between reality (*res*) and language (*verba*). In the final analysis, rhetorical strategies could be used to unite any body of ideas in a playful discursive enterprise, or in a serious philosophical endeavor that invited further questioning.

The Baconian experiential method thus calls for a regulative, heuristic framework for research, rather than a systematic program for the exploitation of nature. Given the complexities of his scheme, one can readily challenge the traditional interpretation that sees instrumentalization and objectification as the two most important features of Baconian science. Instead, Bacon calls for a harmony between the natural and the artificial (the fictional and the technological); this would prove Bacon anticipating late-twentieth-century preoccupations with the symbiotic relationship between machines (the mechanical) and human life (the biological). This relationship of man and machine can be seen in the growing interest in cyborgs, the continual reliance on electronic media, the use of artifical limbs (prosthesis), even in the current debate over the human genome project.

In his writings on this project, Paul Rabinow offers Donna Haraway and French philosopher François Dagognet as presenting challenges to the conventional critique of a utilitarian use of science and technology.[14] Both Haraway and Dagognet claim that the full exploration of life's potentials is blocked by a naturalism that can be located within Greek (Platonic) thought. For the Greeks, "the artisan or artist imitates that which is—nature. Although man works on nature, he doesn't change it ontologically, because human productions never contain an internal principle of generation."[15] This view posits a sense of the inferiority of human works, the superiority of the biological, and the risks attendant upon artificiality. Bacon, while conventionally described as a naturalist, does not completely endorse this Platonic view of mimesis, arguing instead that nature can be shaped using techniques learned through the making of things.

This way of interpreting Bacon brings to mind the work of Donna Haraway, who offers an alternative reading to late-twentieth-century scientific and technological projects which seem to be governed by such utilitarian ideals. In her 1985 "Manifesto for Cyborgs," she claims that "taking responsibility for the social relations of science and technology means refusing an anti-science metaphysics, a demonology of technology, and so means embracing the skillful task of reconstructing the boundaries of daily life, in partial connection with others, in communication with all of our parts."[16] She celebrates the attempt to call into question organic wholes, such as the poem and the biological organism, and to question "the certainty of what counts as nature—as a source of insight and a promise of innocence.... The cyborg would not recognize the Garden of Eden."[17] So too with Bacon who, in the words of Pierre Gassendi, "seeks support from things (*res*) in completing the cogitation of the intellect."[18]

Although Carolyn Merchant and Paige Dubois would see this as a manipulation of nature, scientists such Evelyn Fox Keller (also writing on the human genome project) hold that this kind of experiential model allows nature to shape the observer who, in turn, reshapes that nature for some purpose, namely genetic health. The moral ambiguities of this urge to manipulate, even for the purpose of a healthier global environment, are obvious in this project to recode human genes. One need only consider German genetic experiments during World War II. The controversies over this project serve to crystallize many of the ambiguities that center on Bacon's advancement of human learning, and makes necessary any study which seeks to understand the political nature of Bacon's contention that "truth and utility are the very things themselves; so works themselves are of greater value and pledges of truth than as comforts of life."

# Reckoning Words

# 1
# Renaissance *Res* and *Verba:* Toward a Poetics of Truth

> The world can never quite look like a picture, but a picture can look like the world. It is not the "innocent eye," however, that can achieve this match but only the inquiring mind that knows how to probe the ambiguities of vision.
> —E. H. Gombrich, *Art and Illusion*

In order to understand the cultural forces that created the *possibility* of a Bacon, I will in this chapter briefly delineate several important preoccupations of what is described by traditional intellectual historians as late medieval and early Renaissance modes of thinking, and which seemed to open the way for modern empirical notions of truth. Using this methodology, my plan to explore the inception and acceptance of early-modern intellectual ideas is not to fix dates, epochs, or labels, but to identify and chart changing mental attitudes which would have a subsequent power to persuade and alter social and political alliances. For example, the fourteenth-century nominalism of William of Ockham is often classified by traditional historians as late medieval, while the contemporaneous humanism of Petrarch is considered early Renaissance. I will bypass these categories by suggesting that humanism and fourteenth-century nominalism developed simultaneously, and that their commonalities, such as a leaning toward particularism rather than universalism, had its roots in a mental attitude that preceded the work of both thinkers. Bacon's *Advancement of Learning* was influenced by the nominalist logic of Ockham and Peter of Spain, as well as the humanist rhetoric of Petrarch, possibly via the poetics of Sidney, Spenser, and the French court poets.

Bacon's goal was to critique the textual assumptions about words and reason that lay behind scholastic and humanist forms of knowing and then to chart a new (middle) course that relied not only (or even primarily) on verbal intellectualism but on technical, practical, and quite literal knowledge about making and constructing *things* (works). As such, his central thesis about the importance of *things* over *words* relied, perhaps ironically, on Renaissance notions of *poesis,* which, according to Sidney, was a *making* of things better than nature brought forth or forms such as never were in nature, "wherein I know not whether by luck or wisdom, we Englishmen have met with the Greeks in calling him a maker: which name, now high and incomparable a title it is, I had rather were known by marking the scope of other sciences than by any partial allegation."[1] Sidney further claims that poesy is an art of imitation "that is to say, a representing, counterfeiting, or figuring forth—to speak metaphorically, a speaking picture—with this end, to teach and delight."[2] He is advocating that poets *make* things, just as visual artists and sculptors make material forms and figures which suggest nature but are not subjected to nature. As such, poets must have as great a practical knowledge about the real world (of nature) and the workings of the body as these other artists; yet, to explore the power of language, they must also be aware of literary tradition.

Similarly, one can identify four important features forming the central focus of Baconian restoration of learning—namely, the power of nature, language, the body and technology (or works) to advance the state of England in an increasingly colonial enterprise. Like Sidney, Bacon was influenced by an increasingly rhetoricized logic that sought to eliminate distinctions between the moral philosopher, the natural philosopher, the mathematician, and the historian. Bacon, however, would privilege not the poet but the (new) natural philosopher who understood nature through a less rigorous deductive logic based instead on a practical knowledge of making things work. In this way, Bacon developed an experiential method based, like Sidney's poetics, on a rhetoricized notion of perception and reasoning. The distinction of *res* and *verba,* subject matter and style, had traditionally been argued in the context of opposition, usually as a confrontation between philosophy and rhetoric.[3] By the seventeenth century, and the work of Francis Bacon, this conflict had become an epistemological battle between things and words, between the natural objects of the external world and the verbal figures used to represent them. Verbal representation would be nothing less than an accurate isomorphic reflection of external reality. This idea of language as *mimesis*, duplicating and copying the ontological order of external things and the universal operations of the mind, can be traced through the medieval period to the triumph of things over words in seventeenth-century experiential science.[4] A. C. Howell effectively argues that, beginning with Bacon in the seventeenth century, a new conception of *res* began to emerge, along with a changing conception of prose style and the proper mode of communicating the serious facts of scientific knowledge. "The term *res*, meaning *subject-matter*, seems to become confused with *res*, meaning *things*, and the tendency to assume that *things* should be expressible in *words*, or conversely, *words* should represent *things*, not metaphysical and abstract concepts."[5]

According to Howell, Bacon showed a marked preference for the definition of the phrase given by Quintilian, a preference that separated him epistemologically from earlier writers of the late sixteenth century who preferred the linguistic ethic of Cicero. While Cicero discussed *res et verba* using a broader conception of a plain or literal style (*brevitas*) versus a more copious and metaphorical style (*copia*), Quintilian claimed that one style was more appropriate than another: "I would have the orator, while careful in his choice of words, be even more concerned about his subject matter. For, as a rule, the best words are essentially suggested by the subject matter and are discovered by their own intrinsic light."[6] In his *De Augmentis Scientiarum*, Bacon uses the same phrase as Quintilian to describe the proper way of joining words and things, both in the mind of the reader, the receiver of knowledge, and on the written page intended for the secondary reader: "Curam ergo verborum rerum."

Later, Hobbes, in his *Leviathan* of 1651, would persuade the reader to use plain and simple language, utilizing only those names or words whose meaning has been authorized or guaranteed by external facts or objects. Words, he claims, are wise men's counters; they do but reckon by them. Since there is no reasoning without speech, a wise and prudent seeker after knowledge must first demonstrate the logic among things, then register that cognition by the use of marks or signs—that is, words that act as notes of remembrance, signifying also the connection and order of every conception, desire, fear, and passion. The primary use of speech is to transfer our mental discourse, which is only matter in motion impressed on the mind via the faculties of sense, into arbitrarily defined marks and signs. These signs help the memory or recollection, showing to others that knowledge which we have attained. This process inverts the usual mode of logic based on syllogistic reasoning. In a syllogism, if one can demonstrate the linguistic and relational logic within the three parts of a syllogism, one has automatically demonstrated (or proven) the truth of the subject matter. The logic of things is determined by linguistic necessity.

To reason properly, Hobbes continues, one must start with the settled signification or definition of names, and proceed from one consequence to another. Scholars should not take up conclusions on the trust of authors; rather, they should investigate first significations of words, laboring after meaning rather than simply believing. Reason, unlike inborn sense or memory, is attained by industry, "first in apt imposing of names; secondly by getting a good and orderly method in proceeding from the elements, which are names, to assertions made by connection of one of them to another . . . till we come to a knowledge of all the consequences of names appertaining to the subject in hand; and that is it, men call SCIENCE."[7] Words and things are distinct and separate; words are markers or signs whereby the "fancying" of the object by mans's five senses is set into the memory. This is a primary process of man. Reasoning or reckoning is the secondary process, achieved by labor, whereby correct or apt names are imposed and inapt names discarded; connections are made between these correct elements, and a general assumption is maintained. This latter event is called science. In this contractual mode of language, words are commodities of exchange, marking an extra- or prelinguistic conceptualization of reality, which, for Hobbes, consists in the material objects of the natural, external world.

This notion of things as material artifacts represents a refinement and reduction of the classical notion of *res* as subject matter, encompassing material objects as well as emotional and spiritual states. *Res*, for Plato, referred primarily to ideas as extralinguistic realities apprehended by the mind and then translated into a conceivable form; *verba* was that form translated into speech. In the classic mimetic model, these words copied or duplicated the objects of thought. According to G. F. Else, in his study of pre-Socratic notions of imitation, Plato radically redefines *mimesis* as a direct representation that separates the original from the copy.[8] Before Plato, *mimesis* was used by the pre-Socratics to mean a mimed performance, usually in connection with a Dionysian cult drama or musical performance. My contention is that Bacon, as well as Sidney, followed this pre-Socratic view of *mimesis,* which relied less on an intellectualization of nature and more on an understanding of the physical possibilities and limitations of the body in understanding and representing nature. I must therefore delineate how and why Plato rejected the pre-Socratic philosophers in order to establish a science of imitation premised on memory and an intellectualism that denied the impulses of nature and the body. For Baconian science to develop in the seventeenth century, this Platonic and Aristotelian intellectualism had to be effectively denounced as a faulty mode of knowing about the natural world and then replaced by a new experiential model of natural imitation that relied on the ancient makers of myth, fable, and poetry.

Imitation—as developed by the dramatists Aristophanes, Euripides, Herodotus, Thucydides, and Democritus—involved the manipulation of the living voice, gesture, dress, and action. A secondary sense extended *mimesis* to include *techne,* or the skilled reenactment of a gesture. Aeschylus would often employ an animated doll, while Euripides used embroidered figures or effigies of a person or thing in material form (*mimema*). For these pre-Socratic poets, the underlying assumption of mimetic performance demanded that the actor become like someone else, to mimic the habits and gestures of another dramatic performer. *Mimesis,* in this scheme, becomes the emotional *effect* of the performance, not the actual actors or the performance itself. The members of the audience, via the performance, must assimilate their souls to the character of the person or life being imitated.[9] This illusion of psychological identity, known as *apate* (trickery and deception), enhanced the entertainment value of the performance and gave to tragedy, the usual vehicle of *mimesis,* the added prestige of complexity and fruitful obscurity.

The combination of paradox and metaphoric signification used effectively by Homer, Hesiod, and the later tragedians hints at the mysteries that lie concealed behind the veil of appearances. A manufactured *apate* was to be the goal of every artist, and every viewer must knowingly submit to it. That an audience would gain by this deceit, disguise, or misrepresentation is a central presupposition of myth and fiction. The importance of oracles, riddles and enigmas to the pre-Socratic poets and thinkers testifies that essential realities cannot be conceived or expressed using ordinary speech habits. The complete depths of the *logos* cannot be penetrated without recourse to these deliberate ambiguities and multiple levels of meaning. In this spirit, Heraclitus writes "the lord whose oracle is in Delphi neither declares nor conceals but gives a sign."[10]

In this brief fragment, typical of his enigmatic and elliptical style, Heraclitus alludes to a process of attaining knowledge that bypasses the ordinary avenues of speech and reason. Oracles require reflection and interpretation on the part of the listener; ordinary language is insufficient to communicate certain essential realities. Instead, the listener is forced to question and puzzle out an answer, ultimately provoking the proper response to a situation. Oedipus, to his misfortune, neglects to reflect on or question the identity of the stranger on the road, and this lack of reflection, this ultimate and fatal blindness, leads finally to his physical and spiritual disfigurement. Heraclitus recognizes the potency of this mode of communication, suggesting that riddles, enigmas (that is, broken knowledge) form a powerful avenue through which the *logos* can be penetrated.

This notion of fruitful trickery and riddling—indeed, the deliberate falsification of information—does not have its origin in Heraclitus, although he is in fact one of its most illustrious practitioners.[11] What the ancients recognized in Heraclitus was his radical reformulation of the question concerning words (that is, names) and things. Once the magic identity of *res* and *verba* was finally denounced, the problem confronting poets and thinkers was the nature and correspondence between reality and formal expression. According to Eric Havelock, in the educational culture of Plato, poetry, as a means of expressing truths about reality, held a central, absolute position.[12] Poetry (epic, lyric, and dramatic) represented an encyclopedia of knowledge, giving access to all sorts of useful information on ethics, politics, history, and technology. The epic poet, the *rhapsode* in his role as prophet, teacher, and public entertainer, inhabited a role much like that of the Heraclitean oracle; he neither declared nor concealed, but gave a sign. These signs usually took the form of myth and fable, retelling the exploits of human and divine figures wrestling to understand and confront their particular destinies.

As Havelock acknowledges, poetry occupied a place in the pre-Socratic cultural world that modern commentators can only slightly, and with difficulty, comprehend. Havelock himself gives serious consideration to the reasons behind Plato's utter denunciation of poets.

Surely, given his own eloquence and lyric grace, Plato would not banish the poets simply by virtue of their seductive rhythms and lyric harmonies. Rather, Havelock writes, Plato's target seems to be precisely the poetic experience as such—that is, the method by which poets communicate knowledge to the audience.[13] The key to this method is, of course, *mimesis*. *Mimesis*, or imitation, is an essential component of that relationship, so central to Heraclitus, between words and things. Imitation, for Plato, is an act of composition used by poets to create a story; it is the performance by an actor—or a word, phrase, or sentence—that is the embodiment of that central act of creation. It is this dramatic and re-creative performance that Plato most strongly objects to—in actors and in words. Similarly, *copia*, meaning either a copy and a model, or suggesting abundance, variety, and strength, is described in negative terms:

> no one man can imitate many things as well as he would imitate a single one . . . if they imitate at all, they should imitate from youth upward only those characters which are suitable to their profession—the courageous, temperate, holy, and free; but they should not depict or be skillful at imitating any kind of illiberality or baseness, lest from imitation they should come to be what they imitate. Did you never observe how imitations, beginning in early youth and continuing far into life, at length grow into habits and become a second nature, affecting body, voice, and mind?[14]

What Plato has in mind here is the effective censorship of the imitative arts in the service of ethical and virtuous behavior. One actor playing many different roles—both tragic and comic, virtuous and villainous—is in danger of losing his own unique character and goodness. In the same way, the service of language to perform numerous parts with one word debases the unique, clear, and unambiguous meaning or definition that any word or phrase possesses. Plato, in the *Republic* if not in all his writings, is concerned about this propensity for falsehood and deceit which is a natural part of the imitative arts.

As I have already noted, *mimesis* for the pre-Socratic poets and thinkers involves the actual miming of looks, actions, and utterances through musical voice, manipulating the living voice, gesture, and actions to create an illusion of reality. This work of falsification inherent in any mimed performance is termed by the ancients as *apate*, the fruitful trickery of Gorgias. According the Rosenmeyer, this pre-Socratic thinker, poet, and comedian exploited the use of doubleness in tragedy, indeed in all dramatic mimed performance, to "rationalize the effects of illusionism in tragedy as a contrived *apate* which it is the business of the artist to achieve and equally of the audience to submit to."[15] That gap which naturally exists between performance and reality, present in the creative act of the performer/composer and comprehended by the audience, can be fruitfully manipulated by the composer to comment on fundamental philosophical questions. *Apate,* the trickery, deceit, and illusion present within mimed drama and fiction, is possible only when the magic and performative correspondence of words and things has been removed, when the culture, by way of the poet, is able to ask whether reality is properly communicable.

Plato, of course, was concerned with similar philosophic questions concerning the expressibility of Ideas. For this reason, primarily, he renounces the claims made by dramatic poets to address these issues. Tragic drama, utilizing ancient myths and fables, is not the proper vehicle for the transmission of truth precisely because of the "double nature" inherent in dramatic representation, founded on the ambiguous and multidimensional aspects of *mimesis.* The gap between fiction and truth must be negotiated, if at all acknowledged, by the philosopher, and not the imitative poets. For poets are dangerously close to that "lie of the soul" which is the abomination of mankind. Although the lie in words (that is, the tales of mythology) "is only a kind of imitation and shadowy image of a previous affection of the soul, not pure unadulterated falsehood . . . [and] is in certain cases useful and not hateful," the guardians of the Republic should protect themselves from any unnecessary alliance with poets and dramatic actors.

All poetical imitations are ruinous to the understanding of the hearers by addressing that faculty in man which is most shallow, base, and frivolous. The poet holds up a mirror and, by turning it round and round, presents the sun, moon, and stars, as well as all the animals and plants on earth. By gazing in this mirror, however, the audience sees only appearances or, rather, the semblance of the original appearances (the things of the world) which themselves give form to some transcendent Idea. Thus is the poet, and thereby the audience, thrice removed from true existence. Imitation, then, is only a kind of play, a sport or magic representing objects at a distance and expressing truth only indistinctly, through a veil or curtain.

> The body which is large when seen near, appears small when seen at a distance . . . and the same objects appear straight when looked at out of the

water, and crooked when in the water; and the concave becomes convex, owing to the illusion about colours to which the sight is liable. Thus every sort of confusion is revealed within us; and this is that weakness of the human mind on which the art of conjuring and deceiving by light and shadow and other ingenious devices imposes, having an effect upon us like magic.[16]

Of course, the arts of measuring, numbering, and weighing come to the rescue of human understanding. Reality, which is rational, scientific, and logical, must likewise be expressed using definitions that are clear and unambiguous, with no type of refracting screen between external objects and the perceiving eye that would disguise and distort that reality. Not only is the viewer deceived, believing in illusion rather than truth, but that viewer must submit wholly to the trickery of the poet, acknowledging the false to be true and sympathetically identifying with an experience that refers only indirectly to reality. Unlike Aristotle, who at least sees some kind of cathartic usefulness in viewing the agonistic wrestlings of mimetic performances, Plato, in *The Republic* at least, sees the lies of imitative artists as bred by ignorance and self-indulgence. The only poetry he will admit into his Republic as a means of teaching philosophical and ethical truths are the hymns to the gods and to famous (virtuous) men. Similarly, simple narrative is allowable as long as the composer does not indulge in dramatic re-creations of speeches, thereby transforming himself into another dramatic character.

Notwithstanding these complaints, Plato offers a reconciliation between that ancient quarrel between philosophy and poetry. The gracious sister art of imitation is allowed to return from exile and make a defense of herself in lyrical or other meter. If she can thus prove herself useful to the state, and not only as a source of delightful entertainment, the guardians of the Republic will embrace her as a lover. By issuing a philosophic, reasoned defense of herself, the arts of oral discourse will cease aimlessly to seduce the audience and, following the example of philosophy, will fill that gap between performance and audience with an interpretive and defensive text. As Havelock writes, Plato

> asks of men that . . . they should examine (their) experience and rearrange it, that they should think about what they say, instead of just saying it. And they should separate themselves from it instead of identifying

with it; they themselves should become the "subject" who stands apart from the "object" and reconsiders it and analyses it and evaluates it, instead of just "imitating" it.[17]

In other words, Plato replaces *mimesis* as a communicative mode with *episteme,* which implies science and logic.

In this new Republic, Plato seeks to replace an oral educational system with a pedagogical scheme based on rational and scientific premises, premises that rely on nonpoetic strategies of memorization. Oral instruction in the early Greek *paideia,* from Homer to Hesiod through the tragedians to the sophists, was based on memorization skills that relied not on knowledge of things but on vivid imagery linked to rhythmic and musical verbal patterning. Havelock defines the manner in which a semiliterate culture transmits and preserves its political and social habits, mores, and institutions. A collective social memory is maintained and transmitted from one generation to another through easily remembered verbal and metrical patterning. Whereas a single prosaic directive is impossible to remember fully intact, a vivid image, fixed to a particular metrical sound, might be stored more or less successfully from generation to generation. In addition, memorization must be constant, and in this capacity it is most effective when performed in some form of public ritual, in the theater or the marketplace.

> The community has to enter into an unconscious conspiracy with itself to keep the tradition alive, to reinforce it in a collective memory of a society where collective memory is only the sum of individuals' memories, and these have continually to be recharged at all age levels. Hence Plato's *mimesis,* when it confuses the poet's situation with the actor's, and both of these with the situation of the student in class and the adult in recreation, is faithful to the facts.[18]

There is immense pressure for the composer, as creator and performer, to transform himself into the text, to identify completely with the representative character and figure. It is the mimetic process, then, the reliving of all experiences in the memory, that Plato opposes strongly. A more effective alternative would be a rational, scientific correspondence between words and things, which would be useful in analysis and in the arrangement and classification of experience. Such an isomorphic correspondence cannot imitate action in the same dramatic way, asking the viewer to believe that which is false or contrary to reason. It is the drama of oral performance that Plato objects to, which serves as a refracting screen, distorting reality and leading the mind away from knowledge of what things really are. With Plato, then, the mimetic correspondence between words and things is articulated not in terms of a dramatic performance that calls attention to itself, rather than to the original action being mimed. Instead, *mimesis* becomes a matter of direct representation separating the original from the copy. The verbal performance must be understood (or read) only as a guide to that original action.

Among twentieth-century language theorists, Paul Ricoeur, in his *Rule of Metaphor,* disputes this notion by suggesting that nature makes room for imitation in abundance, an imitation that exists autonomously, yet still with reference to reality. He argues that poetic *mimesis* should be understood not as a copy, but a redescription that uses extended metaphor just as scientific *mimesis* uses material models, in a heuristic logic of discovery, to teach something relevant and truthful about things of the world. Feeling is transformed (metaphorized) into myth, which opens and discovers the world. Ricoeur begins by defining metaphor as a trope of resemblance, constituting displacement and substitution. He further uses the idea of the scientific model to demonstrate the heuristic function of poetic discourse. That is to say, he sees in the poetic statement a logic of discovery, whereby the poet uses his mastery of metaphor to show both new possibilities and new worlds. Ricoeur claims that metaphor is to poetry what the model is to the scientific imagination. Like metaphor, the vehicle of substitution, the model is a type of scientific fiction, set aside and resembling a reality that is too large or indistinct to comprehend without such a concrete artifact. The imagination must be taught to "see as," borrowing an expression from Wittgenstein: to see the similar within the dissimilar.

In discussing his rule of metaphor, Ricoeur acknowledges that modern interpreters of *mimesis* think they understand the term better than they actually do. Indeed, critics of literature seem generally hostile to *mimesis,* and chastise Aristotle as an enemy to creative and imaginative literature. Ricoeur, however, takes great pains to note that the Aristotelian *mimesis* in language cannot simply be reproduced to a copy of nature.[19] Meaning, for Aristotle, is not the end result of dialectical analysis and a comparison of resemblances. Rather, for Aristotle, "there is *mimesis* only where there is a 'making.'"[20] In other words, imitation refers to something that is not, in a primary sense, natural; it is artificial and relies on *techne,* the technical competence of the craftsman.

*Mimesis,* then, for Aristotle and subsequently for the history of Western literature, is an object that is made using the technical expertise of the artist, the maker. *Mimesis* must have some referential function vis-à-vis reality. The poem must submit to reality, yet it must still admit a creative component. *Muthos,* the structure of the plot, allows the poem to be a true imitation, rather than a mere duplication of reality. This definition of Aristotelian *mimesis,* Ricoeur says, "holds together this closeness to human reality and the far-ranging flight of fable-making" (*Rule of Metaphor,* 39). A work of art has an initial reference to reality which, in the process of imitating, becomes a virtual mode of reference. Real reference is substituted or replaced by reference to a virtual world, a virtual life of possibilities.

For Aristotle, the imitation of human action must be ennobling and elevating, imitating not what is but what should be in relation to what is. In this sense, the imitation is not a copy or a counterfeit, but an original creation. *Mimesis,* then, is the epistemological counterpart to persuasion in rhetoric, for *mimesis* aims to convince the reader, through "affective" pleasure, of a potentially elevated life, one which closely resembles (imitates) a possible life in the real world. This virtual life is, of course, impossible without first the reference to, and the suspension of, actual life. According to Ricoeur, this interpretation of *mimesis* gives to the rule of metaphor an added weight and significance. Metaphor is itself an imitation of an essential principle of human action, both real and linguistic.

Because metaphor involves a transference, from one state to another, and a deviance and a resemblance in relation to ordinary usage, the metaphorical process enables *poesis* to occur. Metaphor is at the service of poeticizing.

> Considered formally, metaphor as a deviation represents nothing but a difference in meaning. Related to our actions at best, it takes part in the double tension that characterizes this imitation: submission to reality and fabulous invention, unaltering representation and ennobling elevation. This double tension constitutes the referential function of metaphor in poetry. Abstracted from this referential function, metaphor plays itself out in substitution and dissipates itself in ornamentation; allowed to run free, it loses itself in language games. (*Rule of Metaphor,* 40)

## CHAPTER 1: RENAISSANCE *RES* AND *VERBA*

These language games can clearly be seen in the virtuoso verbal performances of the structuralists, those who deny that literary language, the rule of metaphor, has any initial relationship to reality. Reference becomes the signified, just another sign in a network of verbal relationships. Equally, the use of metaphor as mere substitution and ornamentation is also misleading. The "affective" power of metaphor is thereby denied in the attempt to paraphrase and clarify in reference to a particular meaning located outside the text. The proper balance can be restored if we return to Aristotle's definition of imitation, *mimesis,* as referring to reality, but not restricted by it. The poem imitates the full range of human action, found both in actual life and purely in the imagination of the poet. For the poetic craftsman, the range of possibilities is endless. The poet is limited only by the zodiac of wit. Because any poem is detached from the specifics of historical and political reality, the critic, or reader, need not compare the appearance (the imitation or resemblance) with the real in order to discover meaning or truth. The poem becomes an autonomous entity which can be analyzed in terms of its internal structure, *and* its effectiveness in pleasing the audience with its unity and order, its ability to signify active reality.

Ricoeur tells his reader that *mimesis* itself reminds us that no discourse can ever be detached from "our belonging in a world," to the material conditions of intellectual life. The critic, or interpreter, must develop a double sense of discourse, with one part of the sensibilities fixed on the images and their associative referents in the text, while the other remains on the human experience that informs both author and work. The text itself is a creative dialogue between the different elements that exist in the text's arena of discourse. Meaning is created by the reader using both interior criteria and reference to the allusions to outside reality. Finally, Ricoeur moves a step beyond Aristotle, saying that the rule of metaphor in *mimesis* betters nature by not merely describing reality but presenting all things "as in act," as lively and expressive. Ricoeur mimics Heraclitus here, suggesting that poetic discourse, *mimesis,* is more capable than natural and ordinary discourse of presenting and imitating the Idea as Act that lies behind the things of nature. Poetry can mimic the active patterning of the world, a patterning discerned with difficulty by the nonpoetic observer of nature, who sees only the partiality of process. Ricoeur thus relates the goal of *mimesis,* or "signifying things in act," to a logic of discovery whereby ontological concepts are not asserted but explored and questioned through the correlative interchange of potency (potentiality) and actuality. When the poet "signifies things in act" he or she is seeing things as actions. The reader must then ask how this textual act must be interpreted, how the puzzle or riddle of the discourse can be solved. The reader must become assimilated with the text, participating in the emotional and intellectual fabric embedded within the linguistic structure of the literary work.

Many linguists would go even further, asserting that the metaphorical process is essential in the formulation of scientific ideas. Indeed, grasping a metaphor requires an intuitive mental process whereby a reader's cognitive faculty creates an inner tension between image and meaning. Max Black and Roger Tourangeau explain that metaphors have two subjects, a principal subject and a secondary subject, both operating in a fixed system of belief. These two domains interact and create a deeper meaning than either one can achieve by itself. Aspects of the secondary subject highlight some features of the primary subject, while suppressing others, thereby allowing the reader of metaphor to grasp the significance of multiple words or multiple things in a single illuminating moment:

> [Metaphors] suggest that beliefs about the one subject must be altered in several ways to fit the other. Agreement between the metaphor's picture and our own, and incongruence between domains seem to contribute to our liking for a metaphor. Disagreement produces one sort of novelty—new beliefs—and incongruence another sort—a new structure for our beliefs. In metaphor, a system of belief gets new life in a foreign land; it takes root among the alien corn.[21]

For Black and Tourangeau, metaphors have a cognitive value on a par with that of literal statements. Indeed, the metaphorical unit possesses not only a literal meaning but another, truer signification that gives an additional flash of insight. Ricoeur adds that a metaphor "suggests, reveals, unconceals the deep structures of reality to which we are related as mortals who are born into this world and who dwell in it for awhile."[22]

Ricoeur does not focus on the issue of the sense of metaphor in its semantic dimension, but instead on "the reference of the metaphorical statement as the power to redescribe reality" (*Rule of Metaphor*, 6). He starts with the distinction made by Frege between *Sinn* (sense) and *Bedeutung* (reference or denotation). Sense is what the metaphorical statement actually states; reference is the extended interpretation of that statement, or "that about which the sense is stated" (*Rule of Metaphor,* 217). It is a commonplace, Ricoeur implies, to believe that for certain texts—that is, literary—this two-tiered system of sense and referent ceases to function.

In discussing this misleading, two-tiered system, Ricoeur takes as his example Northrope Frye, who divorces poetic language from scientific language (informative or didactic discourse) which includes history. Poetry is not imitating a historical reality with rhetorical ornamentation; it is not a beautified or eloquent copy of a real or historical event. Instead, poetry has a centripetal or internal movement not toward things but toward other words. Whereas Frege describes only scientific statements, Frye argues for the autonomy of literary discourses. In these literary texts (the poem, the novel), the relationship of sense to referent is suspended. Indeed, this kind of discourse has no denotations, only connotations. The meaning of the poem lies simply in the decoding of textual patterns. These patterns, this web of relationships, constitutes the world of the poem; its arrangement, its genre, its style suggest the "affective" mood of the discourse and engenders an emotional and aesthetic response from the reader.

In contrast, the language of science has a centrifugal, or outward, movement which takes the reader outside the discourse, from words to things. This means that there is a less tenuous relationship between the sentences in the text, and those particular things the sentences are referring to. In a work of literary significance, centripetal, or internal, movement of words "presses towards the broader verbal configurations that constitute the literary work in its totality" (*Rule of Metaphor,* 225). Poetry—indeed, all works of literature—functions like mathematics, ignoring reality and relying solely on its own hypothetical postulates. In other words, meaning in literature is literal; it says what it says, and nothing else. Significance, or meaning, is generated autonomously within the text itself. Literary language, in comparison to scientific or historical discourse, is characterized as "opaque" rather than "transparent." "Opaque" language is defined as "'discourse so well covered with patterns and figures that no vision can penetrate behind it. This would be a language that does not point to any reality, a language satisfied in itself.'"[23]

Language theory, it seems, cannot remove itself from this dichotomous paradigm of the verifiable (transparent) discourse and the unverifiable (opaque) discourse. This dichotomy—articulated originally by Plato, whereby facts are set against emotions, the verifiable against the unverifiable, things against words—is as unsatisfactory as it is difficult to escape. More than one theorist has expressed dissatisfaction with the notion that the literary text is somehow separated from the realm of verifiable facts by an emotional and aesthetic barrier. Literary discourse, like scientific or historical discourse, must exhibit some awareness of things, of the specific and particular world that created it and to which it must refer—if only obliquely, in a deviant, stylized manner. These theorists—which include Jean Cohen, Mikel Dufrenne, and Paul Ricoeur—ask how to open up language "towards a poetics of things."

The question these critics are asking is how poetic discourse relates to the reality it is meant to describe—to the world "out there," as it were. Does poetry represent reality at all? Is "reality" one of the possible constructs that a poem is imitating or miming? These questions have become necessary due to the challenge of structuralism, which dares critics to ignore the specific historical context the poetic artifact might be addressing. Structuralists follow, or combine, the practices of Frye and Ferdinand de Saussure, the founder of structural linguistics. Like Frye, Saussure displays a distaste for the actual speech acts of ordinary life, what he calls *parole,* choosing instead to focus on *langue,* the grammatical structure that underlies any and all statements. *Langue* is "a system of interdependent terms in which the value of each term results solely from the simultaneous presence of the others."[24]

Jonathan Culler associates or relates *langue* to the competence exhibited by the speaker, reader, writer, or critic in any language—namely, the competence to articulate the system of rules or norms of any language system.[25] In this way, the meaning of a statement or utterance is not to be found in an object or occurrence outside the statement, to which the statement actually refers, but inside the series of utterances. This kind of criticism becomes a meta-language, where the critic uses his or her knowledge of grammar to describe (or re-describe) the meaning of the text. Structuralists, following Saussure, are not interested in the real objects of the words used in any text. When discussing the ink marks c-a-t, the critic-linguist uses brackets to refer to the animal in question, "cat," and no brackets when referring to the word/thing found in the text. The reader understands the meaning of *cat* only in relation to the other associative references in the text.

By thus ignoring the things of reality, critics transform words into things, into artifacts and icons. Structuralist critics such as Todorov and Saussure tend not to be interested in what particular "thought" or "idea" a text represents; reference is suspended and language is satisfied in itself. *Res* becomes not the referent but the signified, and this signified might have little connection with external reality. The relation between word and thing, sign and referent, becomes negligible and arbitrary. Even when the literary work obviously describes some external reality, for the critic, the text remains concerned primarily with its own grammatical construction.

In addition, the interpretation of external reality is closely linked—indeed, dependent on—the systems of language inherited by any culture. Meaning is created, and constituted, by the language itself, by the "script" that men and women have at their disposal. Indeed, we are the very products of the language we share and participate in. No longer can *res* and *verba* be seen as inhabiting two separate realms linked by a natural bond of reflection, reference, and representation. Reality is not represented or reflected by language, but created by it.

Surely, when reading a poem, novel, or essay, the reader is constantly reminded of actions and other phenomena that occur in the ordinary world inhabited by thinking and feeling people. Yet, it is just this kind of mental "experience" that the structuralists subordinate, if not directly deny. An alternative theory, one that accommodates these objections, is proposed by speech-act theorists.[26] In their scheme, language consists of concrete statements (utterances) directed outward, usually to a particular group or situation. The "linguistic community" that exists outside the boundaries of the text (but not outside the discourse) is a "heterogeneous society composed of many conflicting interests. . . . It (is) not simply a matter of asking 'what the sign meant,' but of investigating its varied history, as conflicting social groups, classes, individuals and discourses sought to appropriate it and imbue it with their own meanings."[27]

In emphasizing the persuasive aspects of any discourse, speech-act theory recalls the traditional arts of rhetoric outlined by Aristotle and Cicero. The arena of discourse appears suspiciously similar to the law courts and tribunals wherein the ancient art of rhetoric was practiced. And, like those ancient rhetoricians, modern-day speech-act theorists must analyze the *inventio* of the speech act, namely the subject matter; the *dispositio* and arrangement, or the tactics used by the auditor to create a stunning effect; the exploitation by the auditor of the collective memory of his audience; the elocution and style of the speech act, how the sound, arrangement, and rhythm of certain words create any emotional response so that the rhetorical end is achieved in the saying of the speech; and finally, when possible, the delivery of the speech, the gestures and posture of the speaker. In such a way, a speech act does not simply describe the external world, an extralinguistic event, but is an event in itself.

In its nature as performance or drama, the speech act can, and should, create actions, not merely describe them. The meaning of any speech act relies on the set of conventions shared by both auditor and listener. The words of the wedding ceremony have a binding effect on the participants due to the conventions which the words evoke and through which the speaker and the bridal pair participate. In this sense, the discourse, the words and their associative conventions, constitutes the external reality, while the main players, when hearing and speaking the appropriate words, are affected emotionally by it.

The premise of speech-act theory has much in common with the suppositions behind the dramatic, mimetic performances of the pre-Socratics. Both emphasize the necessity of audience participation in assimilating the mimetic act to their own personalities. This kind of intense identification, which so offended Plato, is the primary cause of his banishment of the poets from his Republic, and of his denunciation of mimetic and metaphorical "performances" in the transmission and articulation of truth. The only piece of information a metaphor can convey is an absurd truth or, more usually, a patent falsehood.[28] This kind of reasoning has parallels in the mid-seventeenth-century distrust of figurative language. Thomas Sprat echoes the assertions that metaphors can express only an absurd truth. In Section XX of his Preface to the *History of the Royal Society* (1667), Sprat suggests that "eloquence ought to be banish'd out of all civil Societies as a thing fatal to Peace and good manners."[29] Although a peaceful and well-mannered man can certainly use the ornaments of eloquence to describe "Goodness, Honesty, Obedience in larger, fairer and more moving Images, to represent Truth cloth'd with Bodies, and to bring Knowledge back again to our very senses, from when it was at first deriv'd to our understanding," a wicked man can use the same figures to the opposite effect, and armed with malice, upset virtue and right reason. At their worst, these tropes are "in open defiance against Reason, professing not to hold much correspondence with that but with its Slaves, the Passions; they give the mind a motion too changeable and bewitching to consist with right practice."

Sprat is concerned not so much with the use of rhetorical figures, but with the abuse of such practices. He is aware how writers and speakers often manipulate the ambiguities inherent in metaphor to obtain "general mischiefs, such as the dissention of Christian Princes, the want of practice in Religion, and the like." And he cautions the natural philosopher to avoid excesses, and thus pronounces eloquence and fine speaking to be the most profound enemies of the Royal Society. Brian Vickers points out that this hostility toward the passions and the imagination, and hence toward metaphor, was not an attack on stylistic ornaments as such. Rather, these new scientists, such as Sprat, Hooke, and Boyle, were using rhetoric, or rather antirhetoric, to attack their political enemies, namely, nonconformists, enthusiasts, mystics, and occultists. The Royal Society, as well as the Church of England, wished to remove itself from the taint of superstition and religious passion. The Royal Society, in particular, wished to stress the objectivity and reasonableness of its research. These scientists sought to dissociate themselves with the hermetic and mystical influence of alchemy. Sprat certainly had good cause to insist that the discourse of the new scientists be objective, clear, and precise. Religion, science, and magic were closely interwoven in Sprat's society; there was only a thin line between alchemy and chemistry, astrology and medicine. Certainly, Joseph Glanville still believed in witches, and Isaac Newton experimented in alchemy, as did Robert Boyle.

In the realm of language, Sprat denies any magical connection between words and things, claiming that any search for such a mystical bond can only lead to intellectual and philosophical confusion.[30] Decorum should be observed at all times. Each word in a metaphor, or in any other similar rhetorical figure, should be set in its proper semantic context. The reader, though less astonished by a sparkling surface analogy, must dig deep within each word of the figure and thereby find a relationship that is decorous, accurate, and universally valid and true. The split between rhetoric and philosophy seems complete. Thomas Sprat would long for a mathematical plainness in language where things were in equal number to words. Metaphor, which he himself used regularly, must be eschewed as deceitful, as the devices of tricksters and women. The peace between *res* and *verba* is disturbed by the vague, undefinable concepts specified by metaphorical figures. Ambivalent and equivocal connotations, outside forces of character and delivery, the emotional seductiveness of verbal drama, give to words, to *verba,* a weight and effectiveness that allows for the sometimes progressive, but more often disastrous, competition to occur between the ephemeral *verba* and the concrete *res.*

At its most stimulating, the *res* and *verba* dichotomy allows words to get the upper hand. This occurs most frequently in the work of writers following Augustine, who make the surprising claim that *verba* actually are more concrete and sensual than *res. Res,* or final reality, could refer only to those pure forms or ideas in the intellect. Words, and the concrete and material things of nature that words resemble, could only be a less fine, noble, and elevated tool of discovery. Truth resides in the ideas (things) of pure form. In this scenario, words are judged and scrutinized in terms of their resemblance to these pure intellectual ideas found in the mind. Natural things are analyzed not in terms of their own form and identity, but in their capacity as signs pointing or referring to an extralinguistic reality. Words and things in this world, in this time, are of equal stature in their signaling capacity. Words and things both serve a higher god, a transcendent purpose that is essentially intuitive in nature, beyond both things and words. As a cohesive, conceptual unit, *Res et verba* triggers an intuitive response in the innermost recesses of the mind, discovering a truth lying dormant until that discursive moment. It is this kind of creative moment that brought the world into existence. "In the beginning, was the Word. . . ."

When the religious element is removed from this model, when the authority bestowed by God on *logos,* the Word, is called into question, the mimetic cohesion of *res et verba* is destroyed. *Res* is separated from *verba,* and linguists begin to provide a dualistic model of language based on a strict definition of words as copies of things. That this dualistic model of language, or linguistic performance, has dominated Western intellectual thought up to the twentieth century is a commonplace in linguistics and literary studies. Twentieth-century aesthetics, as both Richard Waswo and E. H. Gombrich suggest, is no longer concerned with the kind of convincing representational accuracy that interested thinkers, poets, and writers who worked within the model provided by Plato.[31] As such, *imitation* is often dismissed as psychologically simple. Waswo, for instance, identifies a more sophisticated (and sophistical) "relational semantics" among Renaissance writers which refuses "to conceive of meaning as reference produced by signs."[32] This referential semantics is reestablished by seventeenth-century scientific writers who determine the priority of things over words. Waswo argues that in the flight from *words* to *things,* God is established as the ultimate magic to assert the connection that constitutes meaning.

> With meaning thus given, as conferred by reference, semantics can proceed to debate the nature of the something else referred to. Theory has no need to question the connection, but merely relocates what it is a connection to.[33]

My aim in the next chapters is to question the notion that seventeenth-century language theory no longer questioned the connection and resemblance between a sign and its signified, but was content to relocate meaning solely within the referent. A reconstruction of the various critical and interpretive strategies that were fighting for sovereignty in the period shows that seventeenth-century theory continued to be troubled by convincing verbal representation in the creation of meaning. Imitation proved to be more than a simple copy or duplication of external reality. What is lost in an oversimplification of *mimesis* is the elusive and illusionary qualities inherent in any mirrored imitation, qualities of trickery and deceit that late Renaissance writers, including Bacon, were quick to exploit and use. Bacon wishes to read the external world like a poem, to represent or insinuate that world using the same method wherein it was originally invented. This method need not eliminate a metaphoric or enigmatical style, one that allows knowledge to be examined and continued. As such, imitation is inseparably linked to originality, creativity, and, more important, the "discovery of the ambiguities of vision."

Accurate representation and a new understanding of scientific *mimesis* occupied both scientists and poets of the period and set the *res* and *verba* debate within the same predicament that troubled sixteenth-century rhetoricians and translators—namely, the problem of translating "meaning" from one text to another. In this semiotic process, several related problems exist concerning the crucial connection between words and their referent. Indeed, what is "meaning" for the scientist translating the "text" of the external world into verbal symbols; for the rhetorician and translator asking how "sense" can be transferred from one text to another; for the language theorist writing on the universal language of Adam; or the poet deciding on the relationship between images in the mind and verbal signs on the page? To answer these questions, Renaissance philologists turned to the ancient books on rhetoric and poetics, finding in them an appealing conception of the text as rhetorical performance, in which "meaning" is caught in a network of contingent historical circumstances.[34] The mimetic process reconciled for them the source of authorized truth and the individuality of the man-made "counterfeit," thereby resolving the conflict between tradition and modernity. *Mimesis* is for the sixteenth- and seventeenth-century writer, following Augustine, a "mode of self-fulfillment, a striving to become what one admires, a personal realization of a possible ideal—not deference, but performance."[35]

# 2
# *Copia Verborum:* The Maker's Knowledge in Renaissance Poetics and Rhetoric

> Only the poet, disdaining to be tied to any such subjection, lifted up with the vigour of his own invention, doth grow in effect another nature, in making things either better than nature bringeth forth, or, quite anew, forms such as never were in nature . . . so as he goeth hand in hand with nature, not enclosed within the narrow warrant of her gifts, but freely ranging only within the zodiac of his own wit.
> —Philip Sidney, *A Defence of Poetry*

IN THE QUOTE TAKEN FROM HIS *DEFENCE OF POETRY*, SIDNEY claims that the wit of the poet is able to create and invent a structure, a pattern and a world that is not limited to a mimetic or isomorphic representation of the natural world. In his case, Sidney adapts Augustine's distinction between truth and eloquence yet inverts the traditional hierarchy so that the verbal acumen of the poet can express a truth that exists in the mind if not, strictly speaking, in the natural world. Whereas Sidney was considered by his contemporaries a Protestant hero and martyr, the twentieth-century reader can detect a growing scepticism in these words concerning the ability of any text, either the book of Scripture or the book of nature, to determine the originality of a poet's wit and discourse. In fact, in the *Defence* itself, Sidney claims that it is the poet's ambition and achievement of secular fame that constitute and guarantee the "truth" value of any statement. The question I pose in this chapter is how Sidney, and other Elizabethan writers, could arrive at this kind of proclamation, given the strict Aristotelian and Augustinian rules of poetics, rhetoric, and logic that had forged his thinking. In answering this question, I would like to consider how the growing restrictions placed on rhetoric and logic in this early-modern period opened a space in which a thinker such as Bacon could posit a new method of reasoning and writing premised not on a closed system using the commonplaces of rhetoric and logic but on a more open and dynamic method using the tools of empirical observation and a growing awareness of the techniques of making and fashioning "things."

I would like to begin this chapter by anatomizing the Baconian concept of science, in order to determine how the late-sixteenth-century study of, and relationship between, philosophy and rhetoric contributed to the development of an experiential method constituted by practical, technical ways of knowing. According to Charles B. Schmitt, "within the Renaissance textbook tradition, the subject was in general divided into four main fields: logic, natural philosophy, metaphysics and moral philosophy."[1] In addition, the trivium course of study for the undergraduate was divided between logic, rhetoric, and grammar, with a growing elision of boundaries between the study of rhetoric and of logic (or dialectic). As I shall with greater detail demonstrate in this chapter, the early European humanist tradition, responding to the educational reforms of Petrarch, Lorenzo Valla, and Rudolph Agricola, centered on criticism of the professional logician, and the abundant use of terministic ways of conceptualizing logical and technical problems in the university curriculum.

As Schmitt notes, "associated with this disparagement of traditional logic instruction is a general commitment on the part of humanist educators to initiating language study with a study of grammar which attends to the subtleties of the Latin language rather than to the terminology and technical niceties which would be required at a later stage if the student were to pursue logical studies into their higher specialist reaches."[2] These humanist pedagogues tended to prefer a general education based on Roman approaches to ratiocination (such as that of Cicero and Quintilian) as opposed to the more technical modes of reasoning found in Aristotle's *Organon*, thereby calling for a more "humanized" and "rhetoricized" logic. Ironically, these humanist reformers were harking back to an older (Roman) practice of studying philosophy in the context of preparation for holding public office, and were reacting against the "moderns" who had developed their discipline as a result of twelfth-century discoveries of Aristotelian (Greek) philosophical texts and which stressed contemplation and intellectualization removed from public and worldly life.

These scholastic philosophers held syllogism as a higher mode of reasoning than topical logic, claiming that the former were to be used in cases of scientific demonstration and certainty, while the latter were to be used for contingent and practical matters, demonstrating "mere" probability. These scholastics treated topics theory as a debased branch of the study of argumentation, claiming that it was better placed with "rhetorical" topics as a part of rhetoric. In contrast, treatises on logic written by humanist educators insisted that topics logic was the core of dialectic, not just a subsidiary part of logic and not an aspect of rhetoric at all. Although one may claim that Agricola and Valla, in so arguing, "rhetoricize" and debase dialectic, and relegate rhetorical topics to the category of "mere" style and grammar, they do succeed in providing an alternative system of reasoning that reveals the (social) limitations in traditional notions of certainty based on syllogistic rigor. By thus providing a rival to syllogistic reasoning, the humanists defended discourses that defied such deductive and demonstrative rigor to be held as special mediums of truth, such as poetry or the Socratic dialogue, and paved the way toward modern empirical notions of truth based on contingency, practicality, technique, and social interaction.

Most important, the humanist emphasis on topics logic, with its stress on *techne* and practicality, led to the development of a logic of language use, as well as an increased awareness of the relationship between *res* (subject matter or things) and *verba* (words or style). This attention to *words* and *things*, which came about through humanist efforts to reform the study of logic and dialectic, meant that "philosophy during the Renaissance was to some extent allied with the humanities, that is, with rhetoric, poetry, and historical and classical scholarship."[3] Humanist philosophy was concerned most frequently with the arts and techniques which allowed some insight into the inner workings of nature.

As such, the scientific revolution of the seventeenth century could not have been realized without the influence of Renaissance humanism as it developed in fourteenth- and fifteenth-century Italy and that it was adapted to fit the conceptual and linguistic needs of sixteenth-century English intellectual life. The work of Michel Foucault and Timothy J. Reiss posit a sudden and radical rupture in the sixteenth century whereby one discourse (of resemblance or verbal patterning) is replaced by a "modern" one (of identity or analytical-referential capabilities) so complete that the moderns were incapable of understanding their medieval predecessors. Similarly, in his work on Copernicus, the German philosopher Hans Blumenberg "sees the development of early modern science as part of the emergence of a new epoch, but one that maintains specific threads of continuity with the medieval past."[4]

In my own work, I would like to adopt many of Foucault and Reiss's assumptions about a discursive and epistemological change occurring in late-sixteenth-century England, yet without positing a radical break with the past. In such a way, I will be following the lead of Blumenberg and, more notably, Paolo Rossi, who wrote several influential works on Bacon in the late 1960s. Rossi criticized that type of historiography

> which believes it can determine the "significance" of cultural movements or thinkers of the past by availing itself of reductive formulas which are then elaborated into a theory of research. . . . Thus historical investigation is reduced to setting off . . . the various "aspects of modernity," and to the routine job of assigning philosophers of the past to the slot that is necessarily due them within a "logical" frame, sketched *a priori*.[5]

Yet, Rossi also criticizes those scholars who, in a polemic with the former group, insist on tracking down the precise historical origins of the various conceptual positions and demonstrating the persistence of traditional cultural themes within the work of so-called revolutionaries, thus denying "all differences of *form* and attitude which profoundly altered and modified the significance of those concepts derived from the medieval tradition." In contrast, a third method—one followed by Rossi, Blumenberg, and myself in this book—attempts to explain how certain key writers used traditional forms of knowing and representation to resolve cultural tensions which these older modes were no longer able to comprehend or negotiate. What will also be important is to explain why the *ideas* selected by subsequent philosophers, and coded as influential to the scientific "revolution" of the seventeenth and eighteenth centuries, cohered to particular *authors,* as though the enterprise needed to justify itself by positing the unique, visionary scientist as a type of creative artist on the order of a Shakespeare or Milton (hence, my title of *Baconian* science inevitably attached to the name of Bacon).

In the case of Copernicus, Blumenberg claims that a crisis in medieval thought led to the *possibility* of a Copernicus and a modern age, in that a tension between the Greek philosophy of nature and the Christian view of an omnipotent God led to a freeing of theoretical curiosity while still encouraging a belief in the dignity of man still in God's image. As such, Blumenberg suggests that "notions of cumulative progress in science and technology are derived directly from early modern experience. There is continuity as well, however, because problems and functions remain from the previous epoch and are 'reoccupied' by new positions and beliefs, once the older notions lose their power to persuade."[7] However, Blumenberg does not explore why the name and figure of Copernicus would be so important as a vehicle to carry these notions to subsequent generations of mathematicians and scientists.

To begin, one must look to see how medieval philosophers were negotiating between the primacy of logic and rhetoric. The distinction between speech and reason preoccupied thinkers throughout the medieval period. Although the Christian Latin writers of the fourth and fifth century A.D. tended to distrust philosophy, the Schoolmen of the thirteenth century, while continuing to pay lip service to rhetoric, tended to stress the importance of logic as an art of thinking over rhetoric as an art of persuasion.[8] This subtle shift was due, no doubt, to the greater availability and popularity of Aristotle's works on logic, translated into Latin and widely circulated after the middle of the twelfth century.[9] However, while the study of logic was demanding more attention in the universities, there existed no real consensus regarding the proper domain of logic, or its distinction from dialectic. There developed an uncertainty over where the common places of dialectic ended and where the more technical aspects of formal or demonstrative logic began.

In the writings of Aristotle, the distinction between certainty and mere probability had been worked through in two works known since the sixth century as the *Organon,* or the instrument of science. The first treatise, the *Categories,* deals with the classification of the different types of predicates (substance, quantity, quality, relation, place, time, situation, state, action, and passion) and tends to be more metaphysical than logical. The *Topics* and its appendix, *De Sophisticis Elenchis,* are concerned with dialectical reasoning, with "reasoning which proceeds from opinions that are generally accepted and opposed to demonstrative reasoning, or reasoning from premises which are true and primary."[10] The third work, *De interpretatione* (*Peri Hermeneia*), deals with the analysis of statements that are opposed to each other, and is thus linked with the concerns over dialectical arguments found in the *Topics*.

The last, and, according to scholars, most mature works on logic are contained in the *Prior Analytics* and *Posterior Analytics*. In these two works, Aristotle claims his subject matter to be demonstration—that is, demonstrative science. Through this investigation of reasoning, Aristotle develops the doctrine of syllogism. Although both demonstrative and dialectic reasoning use the syllogism, there exists a key distinction.[11]

> Thus a syllogistic premise is just the affirmation or denial of something about something else in the way we have described. It will be demonstrative it if is true and derived from the first principles of the science with which it is concerned; while a dialectical premiss is, (a) when one is inquiring, the invitation to make choice between a pair of contradictions, and (b) when one is inferring, the assumption of what is apparent and probable, as has been said in the *Topics* (*Prior Analytics*, book I, chapter 1).

Because of this distinction, Aristotle warns against using dialectic as a method of scientific demonstration. Indeed, he writes,

> the more we try to make either dialectic or rhetoric not what they really are, practical faculties, but sciences, the more we shall inadvertently be destroying their true nature; for we shall be re-fashioning them, and shall be passing into the region of sciences dealing with definite subjects rather than simply with words and forms of reasoning.[12]

In the twelfth century, however, logicians were more impressed with the *De Sophisticis Elenchis* than the *Prior Analytics*. This comes as no surprise, since the former work follows closely upon the *Topics* (which the medievals would have known from the translations of Cicero and, later, Boethius). In Peter of Spain's *Summulae Logicales*, the standard textbook of logic in the Middle Ages, the *Posterior Analytics* is neglected entirely, along with its concern for strict demonstrative reasoning. Instead, Peter of Spain identifies logic, the art of thinking, with dialectic, the study of arguments. Indeed, strictly speaking, the *Summulae* is not at all a manual of logic. "Its object was to prepare the minds of beginners for the dialectical tournaments and disputational examinations of university life."[13] At the beginning of the treatise, Peter argues that the word *dialectic* is derived from Greek words that mean speech (*sermo*) and reason (*ratio*). "Dialectic, that is, logic, is thus the expression of reason in words."[14]

In this sense, Peter differed in his understanding of the function of logic from Aquinas, who offered a clear distinction between mental discourse (logic as a *scientia rationis* where the mind operates upon things), and speech or natural language. For Aquinas, truth was not in things but only in the mind's perception of things ("*dicitur oratio vera, in quantum est signum intellectus veri*").[15] Logical discourse could, in this sense, be called inner discourse. However, because man is a political and social animal, this inner discourse must ultimately be articulated. Aquinas, in his commentary on Aristotle's *De Interpretatione*, writes: "since logic is ordered to obtaining knowledge about things, the signification of vocal sounds, which is immediate to the conceptions of the intellect, is its principal consideration."[16] Aquinas further explains the correspondence of things and their signs, as well as the distinction between natural signs and those instituted artificially.

> [Aristotle] says here that letters are signs, i.e., signs of vocal sounds, and similarly vocal sounds are signs of passions of the soul, but that passions of the soul are *likenesses* of things. This is because a thing is not known by the soul unless there is some likeness of the thing existing either in the sense or in the intellect. Now letters are signs of vocal sounds and vocal sounds of passions in such a way that we do not attend to any idea (*ratio*) of likeness in regard to them but only one of institution, as is the case in regard to many other signs, for example, the trumpet as a sign of war. But in the passions of the soul we have to take into account the idea of a license to the things represented, since passions of the soul designate things naturally, not by institution. (*Peri Hermeneias, comm. Aquinas*, 27)

This suggests that, for the medievals, an understanding of natural objects must be found in the soul rather than, strictly speaking, in nature. Furthermore, of the five species of perfect speech (enunciative, deprecative, imperative, interrogative, and vocative), only the first is able to signify the conceptions of the intellect absolutely. "It follows that all the modes of speech in which the true or false is found are contained under the enunciative, which some call *indicative* or *suppositive*."[17] The remaining four belong to rhetoric or poetics.

> Enunciative speech belongs to the present consideration and for the following reason; this book is ordered firstly to demonstrative science, in which the mind of man is led by an act of reasoning to assent to truth

> from those things that are proper to the thing; to this end the demonstrator uses only enunciative speech which signifies things according as the truth about them is in the mind. The rhetorician and the poet, on the other hand, induce assent to what they intend not only through what is proper to the thing but also through the dispositions of the hearer. Hence, rhetoricians and poets for the most part strive to move their auditors by arousing certain passions in them, as the Philosopher says in his *Rhetorica*. This kind of speech, therefore, which is concerned with the ordination of the hearer toward something, belongs to the consideration of rhetoric or poetics by reason of its intent, but to the consideration of the grammarian as regards a suitable construction of the vocal sounds.
> (*Peri Hermeneias, comm. Aquinas,* 62)

In Aquinas, then, a logic of scientific demonstration dealing with propositions of truth and falsehood was distinguished from a logic of probability that concerned debate or argument, usually conducted in a legal forum. This dialectic, in turn, is distinguished from rhetoric, used to persuade a particular audience of a practical action, and poetics, dealing with a potential or golden world where men and women act in an ideal way. In other words, the study of truth regarding the objects of the natural world would have been closely tied to a certain conception of rhetoric and speech.

These distinctions, however, would have been taught in the theology course, not a natural philosophy or medical course. Here, Aquinas's contemporary, Peter of Spain, proved the more influential thinker and writer. According to Peter of Spain, logic, the art of arts, is more a *scientia sermocinalis,* or verbal science, and there is a greater connection between inner discourse and outward manifestation, or speech and writing. Logic is necessarily the expression of reason in words. This is not surprising, since Peter was writing for the university arts student who would have been concerned with the persuasiveness of his oral disputations. And, as Walter Ong tells us, this logic was more oriented toward medicine than theology.[18] Thus, in contrast to the logic of Aquinas, Peter's tended at the outset to be more practical, as well as being tied to the needs of an experimental natural science.

The *Summulae logicales* was divided into twelve tracts. The first six dealing with the Aristotelian notions of proposition, predicables, predication (the categories), syllogisms, topics (or places), and fallacies, whereas the second six tracts dealt with those things particular to medieval logic—namely, supposition, relative terms, extension, appellation, restriction, and distribution. These last six tracts have come to be known collectively as the *Little Logicals* (*Parva logicalia*), or the Modern Logic (*Logica moderna*). The key to these last six treatises would be *suppositio,* which dealt with terms that "suppose" or "stand for" individual existence. *Suppositio* is distinct from *significatio,* which Peter defines as "the representation, established by convention, of a thing by an utterance" and is the process by which meaning is conventionally imposed on sound to create words that name things.[19] In contrast, supposition dealt with the property of a previously signified term in any proposition. Supposition, then, is the method of analyzing the relationship among a number of signs in any given statement.[20]

> Supposition and signification differ because signification is accomplished through the imposition of a word to signify a thing, while supposition is the acceptance of a term, already significant, in place of something (*pro aliquo*), as when one says, 'man runs,' the term 'man' is taken to stand for Socrates, Plato, and the rest of men. Thus signification is prior to supposition, and they are not the same. The two differ in that signification belongs to the word (*vox*), whereas supposition belongs to the terms (*termini*) already composed of word and signification. Furthermore, signification refers the sign (*signum*) to the thing signified (*signatum*), whereas supposition refers that which stands for something (*supponens*) to the supposit (*suppositum*). (*Summ log.* TR. VI, 6.03)

The theory of supposition is concerned (and connected) with the categories of thought, with the comprehension and extension of the meaning of signs in relation to their subject. According to Ong, *suppositio* led to a quantification of thought similar to modern mathematical logic.[21] Rather than employ mathematical symbols, terminist logicians used language; thus, *suppositio* necessarily included the study of relatives, extension, appellation, restriction, and so on, in order to obtain a greater precision of thought by reducing and restricting the possible additions to meaning that any given word contained. The aim of this terminist logic, with its distinction of *significatio* and *suppositio,* was to construct a referential language whereby conventional speech acts would embody the order of things found in reality. Yet, in the absence of calculus, the terminist logician necessarily had to subordinate words to things, eliminating the historical and cultural meanings attached to terms in order to create an abstract science. Hence, the need continually to restrict and qualify the terms that supposed or stood for some external thing, either a physical object or a conception in the mind. The goal was to create a precise and natural language that would serve as an instrument of philosophizing. Or, as Elsky claims,

> supposition recognizes the dependence of propositions on conventional *vox* but seeks to guard against and correct distortions resulting from idiomatic usage by continually clarifying the referential relationship of words to things. In this sense supposition theory abstracts conventional speech from the conditions of its utterance in places where people actually speak or construct utterances with cultural coordinates.[22]

Another point should be made about Peter's intellectual and epistemological distance from Aquinas. In Peter's logic (or dialectic, since he uses the terms interchangeably), the arts student, while claiming to deal with scientific certainties, actually is concerned with probable argumentation, aimed at producing conviction in the audience. *Ratio,* or reason, is not linked to demonstrable certainty and truth, but to confidence and trust (*fides*) in probable matters. This emphasis on argument, persuasion, and conviction turns Peter's logic into dialectic, understood in connection with rhetoric, as the art of discoursing well. In contrast, Aquinas separates *oratio,* or inner discourse, from *voces,* or outward discourse with its implication of social interaction where the arts of persuasion are needed. He distinguishes between understanding and discursive reason, the latter being an ability found in the human intellectual apparatus "due to the material component in man's cognitional make-up and in the make-up of the reality he is immediately faced with."[23] The discursive process entails a drawing forth of possibilities found either in the external world or in the inner storehouse of memory. This ratiocinative process exists in time, and allows one piece of "truth" to be revealed after another, then ordered or distributed by the mind to complete a full picture of reality. This discursive procedure is strictly regulated by the material things in the world of matter.[24]

However, for Aquinas (perhaps following Augustine), beyond this human intellectual activity lies a more remarkable act of intuitive understanding. Peter's logic does not comprehend this intuitive moment, and stops with a process of reasoning whereby verbal statements (or arguments) are given an artificial precision in order to analyze the structures of reality. Truth and falsehood might be detected by the ratiocinative process, with reason simply analyzing faulty linguistic structure.

Perhaps because of his rigid linguistic categories, Peter of Spain was reviled by subsequent sixteenth-century humanists. Theorists such as Lorenzo Valla, Juan Luis Vives, Rudolph Agricola, and Erasmus argued against the rigidity and artificiality of terminist logic, as well as the attitude toward language displayed by the Schoolmen. The attempt to isolate language from its historical and cultural milieux, which was anathema to these writers, can be seen in their insistence on replacing *oratio* with *sermo*.

Erasmus, for instance, claimed that "verba" had a triple meaning; "first, a word is formed when it is produced by the voice; secondly, it is formed in the heart with no movement of the tongue; and third, when the thing itself which that word signifies is conceived in the mind to be what it is." For Erasmus, a word is first uttered before its significance is conceived in the mind. Conventional usage of language precedes both intuitive comprehension and logical analysis of reality. Or, as Vives claimed, language was not made to conform to the rules; instead, the rules were made to follow the pattern of language. Logic would subsequently determine the truth in any given utterance or statement.[25] Verbal discourse, as it was spoken by common people (by which is meant university men) would be given primacy over logical abstractions.

In his writings on religious rhetoric, Erasmus claims a preference for "sense" (*sensus*) over wisdom (*sapientia*), and the association of sense with voice. S*apientia*, translated as wisdom, had traditionally been associated with the Word, the second person of the Trinity. Sapience, pertaining to the higher faculties of the intellect, had always been given superiority over sense, which has commonly been linked to the physical and bodily faculty of perception and feeling. The OED relates sense to the French *sentire*, to feel, and associates it with the bodily organs, how a witness perceived external objects and changes in the condition of the body.

For Erasmus, the *verbum* of John's Gospel related to bodily sense and to the physical apparatus of the voice; it was stressed in its incarnational possibilities, in the fact that Christ, the Word, and the second person of the Trinity pitched his tent among men. Erasmus minimized the issue of the Word existing purely in the mind of God. In addition, this sense of God related to intuitive knowledge, to instinct, rather than to strict logic or wisdom. Also, sense implied the signification of a word and the discovery of meaning. Humans participate in and with God and nature to an equal extent that God/nature participates with mankind, in a mutual and reciprocal relationship. It is difficult to see, in this tract, which would be more sacred—the Word of God as he conceived or thought of himself, or the written words of Christ uttered to the men and women of first-century Palestine and recorded by the evangelists. The fallen language of man was given a new dignity by the fact that it was used by the incarnate God, then reused by his followers.

Since Erasmus conceived of *oratio* as *sermo*—that is, as dialogue spoken among men and found in the Greek and Latin classical authors—he tended to emphasize verbal structures rather than intellectual abstractions. In his treatise *De ratione studii ac legendi interpretandique auctores,* Erasmus divided words and things into separate categories, whereby the word was held subordinate to the thing signified. "In principle, knowledge as a whole seems to be of two kinds, of things and of words. Knowledge of words comes earlier, but that of things is the more important."[26] Erasmus, however, did not define the proper elements that fit into the clearly separated grouping of *res* and *verba.* Instead, he claimed that things only have a reality or presence by virtue of the language that expressed those things. As Terence Cave aptly describes *res,* it is not prior to *verba,* nor does it exist in some remaining fashion; but "*res* and *verba* slide together to become 'word-things.'"[27]

Erasmus writes, "but some, the 'uninitiated' as the saying goes, while they hurry on to learn about things, neglect a concern for language and, striving after a false economy, incur a very heavy loss. For since things are learnt only by the sounds we attach to them, a person who is not skilled in the force of language is, of necessity, shortsighted, deluded, and unbalanced in his judgment of things as well." Erasmus criticized those writers who boast about passing over mere words in order to concentrate on the matter itself.

The verbal signification attached to any particular subject matter is inexorably linked to the essence of the thing signified. As in Augustine, Erasmus claimed that the sound of a word triggers in the memory an idea of the thing in question, allowing communication to happen. Augustine, however, argued that comprehension of *res* was not entirely a matter of response to a verbal signal. Ideas come into the mind through simple observation of a particular activity or gesture. Erasmus, in contrast, did not claim such a clear separation of language and ideas in the mind; *res* and *verba* continually occupied one domain:

> Having, therefore, developed a pure, if not ornate, skill in language, we must next direct the mind towards an understanding of things. Of course some considerable knowledge of things as well as of words is acquired in passing, from these writers whom we read in order to refine our language, but traditionally almost all knowledge of things is to be sought in the Greek authors. For in short, whence can one draw a draught so pure, so easy, and so delightful as from the very fountain-head?[28]

Many of Erasmus' contemporaries had stressed the dialectical nature of verbal discourse, whereby *res* is conceived as a supra-linguistic entity apprehended by the mind and the senses and subsequently signified by *verba*. Both Melanchthon and especially Agricola, in *De inventione dialectica,* limit themselves to the invention of topics, an exercise they classified as dialectic, rather than rhetoric. As such, Agricola tended to emphasize *copia rerum*. In contrast, Erasmus' treatment of *copia* is remarkable in its neglect of *res* separated from *verba* and its emphasis on verbal tropes. This emphasis might well be understood if Erasmus' aim had simply been verbal play without any ethical bias toward truth, if his goals had been similar to those of the sophists. However, his aim is always to make some claim to truth, not to engage in debate for the sole delight of speaking. It seems, then, that for Erasmus ideas were generated primarily through verbal play. Or, more precisely, verbal play excited and stimulated ideas already present in the "treasurehouse of the mind," ideas presumably put there by prior reading of scripture and nature and through verbal stimulation.

For Erasmus "textual abundance (the extension of the surface) opens up in its turn an indefinite pluralism of possible senses. The intention (will, *sententia*) which was supposed to inform the origin of a text and to guarantee the ultimate resolution of its sensus remains for ever suspended, or submerged, in the flow of words."[29] The only guarantee of truth lay in the text's resemblance to another work, namely the original Latin or Greek text from which the new piece of writing is derived or from the book of nature. As such, with Erasmus, there was still some kind of stable meaning to which verbal expression attached itself, either textual or natural. While certainly concentrating on language as constituting reality, Erasmus never conceived of an infinite storehouse of *words* or their adjunctive *things*. Indeed, for Erasmus, appropriate subject matter must be moral and ethical, found within the confines of a specifically Christian library of ideas. These expressions did not indicate that, for Erasmus, *copia* encouraged and even constituted a "perpetual deferment of sense" (Cave, 111). Rather, verbal variety, *copia verborum,* did not defer "sense" but instead excited it, making it memorable to the reader. "The humanists cared less to construct static, scientific theological systems than to move men to acts of charity, to persuade them to share their faith, to inculcate morality, and to do all things not by means of painful pedagogy but rather of delightful discourse."[30]

By constant rereading, the receiver of knowledge was absorbed by the text and became transparent.[31] "Meaning," created and understood by the affective faculty (the affections), was guaranteed by the intuitive center of the self, which then became identical to the voice at the center of the text. In the case of Erasmian and Augustinian hermeneutics, that textual voice belonged to God or nature. In this system, which combined the rational with the mystical, the reader must submit totally to that text which is nature. That is to say, the reader, by being absorbed and assimilated into the text, was made to resemble, imitate, and copy the life (the text) of Christ. The identity of the reader became lost in the submission to the performance of God, which is mimed by the scriptural text. In this sense, we are not far from the mimed performances of the pre-Socratic dramatists, and fruitful *apate* advocated by Heraclitus.

At the center of Erasmus' theology and pedagogy, then, is the notion that language—indeed, the world—must be "felt" by the reader as an active presence. Truth could not be discovered or discerned without the mediating influence of words, of linguistic artifacts.[32] By analyzing natural works textually, rather than simply allegorically in the tradition of medieval exegesis, Erasmian hermeneutics solved the problem of obtaining a transparent cognition despite the opaque nature of language: The "meaning" of any utterance would be the emotion triggered in the reader. To be intellectually and morally instructed was to be "ravished by eloquence."[33]

The *De Copia* proved to be a successful and popular work in the first three decades of the sixteenth century.[34] The sixteenth century's propensity to keep notebooks and commonplace books, filled with apt quotations from classical authors, suggests that the method proposed by Erasmus was quickly followed. It was not until thirty years later that a logic or a rhetoric was written in English. Thomas Wilson produced his logic, *The Rule of Reason* in 1551, and his *The Arte of Rhetorique* in 1553.[35] Both texts showed the influence of Erasmus in conceptualizing how the human mind could first understand nature and then express nature.

According to Wilson, the building of cities, the maintenance of fellowship among people, the establishment of order and hierarchies would have been impossible had men not first been persuaded by art and eloquence rather than reason. The implicit assumption, of course, is that eloquence, tied to reason, allowed some to dominate over others not so quick-witted or ready-tongued. The manual was intended for the ruling lord, rather than the servant, for men rather than women, for the man who wished to live following his own vocation, worthy of fame and, in Wilson's opinion, taken for half a God.

It is worthwhile to consider the differences between logic and rhetoric, as Wilson conceived them. Rhetoric is said to be "an art to set furthe by utteraunce of wordes, matter at large . . . it is a learned, or rather an artificial declaracion of the mynde, in the handelyng of any cause, called in contencion, that maie through reason largely be discussed." The artificiality, the learnedness, of rhetoric seems to be what separates it from logic. Indeed, both disciplines, in the two treaties, share invention of matter, and disposition of the same. In his logic, Wilson places *judicium* as the first part of logic, whereby the logicians frame things together, knitting words for the purpose. *Inventio* is the second part, consisting "in finding out matter, and searching stuffe agreable to the cause." He defends this order by identifying the art of the logician with that of the carpenter. The carpenter must first know the methods for fashioning his timber before he can understand the nature of his matter.

*Res,* whether subject matter or ideas, is less important to Wilson than the knowledge how to fashion it. "Truth it is that naturally we finde a reason or we beginne to fashion the same." And this fashioning, this framing, consists "in knitting wordes for the purpose accordingly." As one discovers how the argument is put together, one can then go forth and judge which arguments are most appropriate for a given cause. Logic, or dialectic—by plainly and nakedly setting forth "with apt words the summe of thinges by the way of Argumentacion," by defining and dividing the nature of any thing—is able to distinguish truth from falsehood.

The function of the orator was to expound only on matters that would benefit man, and to do so most effectively through a series of questions. The questions were of two sorts, either an infinite question, which was without end, or a definite one, which comprehended some fixed solution. The first type of question dealt with universal things and properly belonged to the logician, who wrote of matters without respect of person, time, or place. For example, a logician might ask if it were better to live single or to marry, to be a courtier or a scholar. (Bacon's method in the *Essays* can be detected here.) In contrast, the orator asked definite questions dealing with specific appointments, places, and people. The orator might question whether it was lawful for the English king to marry a foreigner or one of his own subjects. Yet, Wilson, like the ancients, warned that the orator must always be informed by the logician, as the particular was comprehended by the general. Before determining whether William the Conqueror falsely seized power in England, one must, using the methods of the logician, question whether it is lawful for any man to usurp power.

In his division of *inventio* and *judicium,* Wilson followed the course set by Aristotle in his *Topics,* and later reworked by Cicero, medieval logicians such as Peter of Spain, and the humanist Rudolph Agricola in *De Inventione Dialectica*. Wilson was unique only in placing *judicium* before *inventio*. Like scholastic logicians, he unambiguously understood logic and dialectic to be interchangeable. For Wilson, the distinction was blurred, and the analysis of verbal structures became the means to ascertain demonstrable certainty. As such, Wilson discusses at length the predicates of a proposition or "five common words" (*genus, species, differentia, proprium, accidens*), the ten predicaments (*substantia, quantitas, qualitas, relativa, actio, passio, quando, ubi, situs, habitus*), the rules of definition and division, the rules of propositions, and the four kinds of arguments (syllogism, enthymeme, induction, exemplum).

The accumulated wisdom found in the books used by any given society is stored in the treasure-house of the mind. As W. S. Howell notes,

> subject matter presents fewer difficulties than organization, so far as composition is concerned. A society which takes such an attitude must be by implication a society that is satisfied with its traditional wisdom and knows where to find it. It must be a society that does not stress the virtues of an exhaustive examination of nature so much as the virtues of clarity in form.[36]

This kind of clarity regarding *inventio* and *dispositio*, the discovery and transmission of knowledge, is apparent and unambiguous to Wilson. The issue is far less simple in the epistemology of Sir Philip Sidney. In his *Defence of Poetry*, Sidney concerns himself with the boundary between philosophy and poetry, asking where the one stops, and the other begins.[37] The defense centers on the pervasive tendency to associate poetry with fiction, with the imitated invention of the storyteller, the entertainer. For Plato, the fiction of poetry was both dangerous and vulnerable. The written work, like a painting or piece of sculpture, is wrapped in silence, a static artifact. The significance of the text—that is, the original author's primary intention—remains problematic and ambiguous. The final meaning of the particular text cannot be elicited through a dialectical exchange with the original writer. Unlike the speaker, who can clarify or justify through additional argument, the writer remains forever silent. The written word has little chance of defending itself.

Sidney's task, then, is to defend poetry against attacks made on it by philosophers, thereby accepting the challenge made by Plato in his *Republic* when asking defenders of poetry to step before the tribunal of philosophers and make a lyric plea of admittance. This kind of epideictic, or praise oratory, stands midway between poetry and rhetoric, thereby satisfying the demands of both philosopher and poet. In this strategy, Sidney attests to the place of the poet in the political community, specifically, "the poet's place within the body politic of culture."[38] He aims for his work to have power and authority. Poets, like philosophers, are caught and trapped in a network of relationships that are both political and linguistic.

> ... they (are) all directed to the highest end of the mistress-knowledge, by the Greeks called *architektonike*, which stands (as I think) in the knowledge of a man's self, in the ethic and politic consideration, with the end of well-doing and not of well-knowing only. (*Defence*, 29)

The poet must negotiate a strategy that resolves the problem of teaching and pleasing both a brazen and a blazoned world.

Ferguson notes the narcissistic dimension of Sidney's apology, how the bond of love, the love of self, motivates his love of the subject he is discussing. No disinterested, autonomous activity here. Sidney conflates the role of forensic and epideictic orator, the one defending the innocence of poetry, the other praising the virtuous offices of poetry. As such, he successfully removes the obstructions between a "language of play" and a "language of power." In the fable of the belly, the belly stands for desire and the passions. The philosopher and the historian both tend to belittle the passions, to the detriment of their disciplines. They are unaware that the people must be wooed, seduced, and enticed in order to learn, and to perform virtuous action. This lack of self-awareness is not evident in the poet, simply because of his narcissistic, self-reflective tendencies. The imitative arts, the invention and disposition of persuasive strategies, ensures that the poet understands and reflects on the basic desires and passions of the people he is influencing and teaching. As such, Sidney's focus is not on the essential nature of representation, but on use and utility. He does not, then, advocate either eikastic or fantastic modes of representation. Sidney, unlike Tasso and Plato, does not explore the issue of representation, or the essential nature of poetic imitation. Rather, he more pragmatically concentrates on the intention of the imitator, and the potential abuse of poetic imitation by immoral writers.

Sidney urges that poetry be read in a morally useful way. He continues by asserting that imagination precedes action (contrary to the popular opinion that the reverse is true), and therefore the prudent soldier should be advised by Homer rather than by Aristotle: "honest King Arthur, will never displease a soldier; but the quiddity of *ens* and *prima materia* will hardly agree with a corslet" (*Defence*, 56). This leads Sidney to criticize Plato, the founder and guiding light of the Greek Academy, and the banisher of poets. Philosophers take the divine delightfulness out of poetry. What is left is "the right discerning true points of knowledge." This the philosophers put into method, and condemn their former masters for the trivial delightfulness of the poetry (*Defence*, 58). Yet, it is this "trivial delightfulness" that is truly important, for it entices the reader, the searcher of true knowledge and self-awareness, to continue the struggle for truth.

Poets can imitate only what is there to be imitated, such as the many-fashioned stories of the gods and goddesses. Poetry, then, is equal to philosophy, in that both imitate reality—what is there. In fact, poetry exceeds philosophy in its ability to move by pleasing. Sidney implicitly, and playfully, denies the stability of reality and truth. Since the truth of the world is foolish, why bother with explaining the "likeness" of a foolish reality. At least the poet does not try to affirm or suggest that his imitation is a true resemblance.

Philosophers and historians lie to themselves, and to their audience, when they claim a statement, a precept or an example, is true. Signs can never be identical to things.

> We see we cannot play at chess but that we must give names to our chessmen; and yet, methinks, he were a very partial champion of truth that would say we lied for giving a piece of wood the reverend title of a bishop. (*Defence,* 52)

This kind of quibble by a serious philosopher is an absurdity. There is a kind of universal truthfulness in the poet's playful exploration of the relationship between signs and things. A stable relationship between words and things can never be achieved. Indeed, the writing of poetry shows how arbitrary, and yet still decorous, is the task of assigning words to things.

Sidney answers here the charge that poets are in the business of feigning, lying, and creating fictions and counterfeits rather than any accurate representation of nature. "And therefore, as in history, looking for truth, they may go away full fraught with falsehood, so in poesy, looking but for fiction, they shall use the narration but as an imaginative ground-plot of a profitable invention" (*Defence,* 53). The poet "nothing affirms, and therefore never lieth" and is therefore more honest than either the philosopher or the historian. The poet, in addition, triggers and stimulates further reflection on any given subject by darkly and enigmatically disguising meaning so that profane eyes will not abuse the mysteries of nature.

Feigning is also useful for servants trying to negotiate their own particular power relationship with their masters. Sidney relates the story of Zopyrus, the faithful servant of King Darius:

> Seeing his master long resisted by the rebellious Babylonians, (Zopyrus) feigned himself in extreme disgrace of his king: for verifying of which, he caused his own nose and ears to be cut off, and so flying to the Babylonians, was received, and for his known valour so sure credited, that he did find means to deliver them over to Darius. (*Defence,* 36)

In an oblique and ironic reference to the status of the courtier/poet, Sidney advocates such an "honest dissimulation," which saves the life of the master and thereby brings honor to the servant, albeit at the expense of several body parts.

Although the poet exploits the playful relationship of words to things, conjuring and feigning relationships, resemblances and appearances to promote his own interest, he must still appeal to the reader's critical judgment. In the digression, Sidney shows how poets must "look themselves in an unflattering glass of reason." Whereas poesy is a divine gift, Sidney says, "yet confess I always that as the fertilest ground must be manured, so must the highest-flying wit have a Daedalus to guide him" (*Defence*, 63). Sidney warns that these artificial rules must not be followed with overly strict regularity, lest the "inventor" fall into the errors of vanity, following what Bacon would later call the Idols of the Tribe.

It appears that here Sidney is inhabiting the role of philosopher taking the poet to task for unbridled flights of fancy.[39] However, in keeping with the rest of his defense, Sidney is actually criticizing the practices of philosophers. Philosophers try to pin down the meanings of words, to establish a stable and unambiguous relationship of words to things. In contrast, the poet realizes the futility of this task. Words must always exceed things. All words contain nuances and subtleties of meaning that make an unproblematic and unambiguous assignment to things virtually impossible. Yet, if this is true, how can the poet, or any writer, ensure that the meaning of his or her words are understood in the network of possible meanings?

The poet must include an explanation, a defense, of his own poetry and his reading strategy. Significance does not come from an authorized source beyond the historical text, and is not generated autonomously within the text itself. Instead, meaning and signification are related to the reading process, to the stages of self-awareness through which the poet asks himself and his reader to pass. In the *Exordium*, which begins the text, Sidney relates the tale of the horseman, John Pietro Pugliano. Pugliano's encomium of horsemanship is not in direct discourse; we as readers hear Pugliano speak through the voice of Sidney. Ferguson claims that Pugliano is the poet-maker whose power is morally ambiguous. Therefore, Sidney stands as a mediating figure, putting Pugliano's speech in context and explaining its function to us, the readers. We must therefore take Sidney as our example of an ideal reader/critic. Sidney, of course, resists Pugliano's persuasion by providing his own encomium or defense of poetry, thereby defending his own position within the political body. This bit of storytelling is motivated, then, by self-love and the need to protect himself and his position.

if I have not been a piece of a logician before I came to him, I think he would have persuaded me to have wished myself a horse. But thus much at least with his no few words he drave into me, that self-love is better than any gilding to make that seem gorgeous wherein ourselves be parties. Wherein, if Pugliano's strong affection and weak arguments will not satisfy you, I will give you a nearer example of myself.... (*Defence*, 17–18)

Pugliano's speech leads Sidney toward self-reflection, then self-promotion. The meaning of the speech is not as important as the usefulness of the speech to Sidney as an individual, to create verses and stories of his own, thus ensuring that he does not die without an epitaph.

This idea is brought full circle in the final *Peroration* of the *Defence*. Here, Sidney mocks the reader for listening and believing this speech. "I conjure you all that have had the evil luck to read this ink-wasting toy of mine" (*Defence*, 74). Earlier, Sidney had told his reader that the orator does not "conjure" his audience to believe that a falsehood is true. This inconsistent use of "conjure," and the overly expansive tone of the paragraph, with its repetition of "to believe," leads one to doubt Sidney's exhortation, just as Sidney doubted the encomium of Pugliano at the beginning of the *Defence*.

At this point, it bears remembering how Pugliano praises the horse," the only serviceable courtier without flattery." By analogy, this horse is the poetry, the language, mastered by the poet. The horse is superior to the horseman, its master, because of its very silence, its inability to flatter and promote itself. By the same token, language, the servant of the poet, is silent without the guiding, restraining, and thus interpretive voice of the poet. Language, in the hands of the poet, becomes vulnerable to abuse, simply by its proximity with the rhetorical skill of the flattering courtier/poet. The moral burden is placed on the poet first, and the reader or critic second. The reader, by implication, must be a logician, aided by the tools of philosophy, who is not easily fooled by eloquent praises. Sidney, the reader of Pugliano, writes "if I had not been a piece of a logician *before I came to him,* I think he would have persuaded me to have wished myself a horse" (*Defence*, 17; my italics).

The first lesson to Sidney's readers is to develop the necessary critical skills to read properly a piece of self-promotion and self-flattery. The second lesson is to Sidney the writer: "but thus much at least with his no few words he drave into me, that self-love is better than any gilding to make that seem gorgeous wherein ourselves be parties." An odd lesson for Sidney as poet. He does indeed learn something from Pugliano's praising of himself; not that horsemen are the noblest of men, but that the vigor of feeling ("strong affection") generated by love of self can readily win (or, in Sidney's case, nearly win) even a weak argument. Hence the need for caution in reading any speech or oration.

With this in mind, Sidney begins to contemplate his own vocation as the noblest estate of mankind. Yet, Sidney had just warned the reader (the would-be logician) to be cautious in the face of epideictic eloquence. By analogy, Sidney calls into question his own praise and defense of poetry, as well as the motive for promoting his own unelected vocation. This word *unelected* allows Sidney an avenue of escape. This vocation of poet has been thrust upon Sidney. As such, the charge of flattery or narcissism or self-conceit becomes problematic and ambiguous. His role as servant is charged with duty, laid on him by another as yet unnamed master. Indeed, the exordium ends on an obsequious note, with Sidney begging pardon for any offense he might give in discharging his duty to defend "poor poetry." Certainly by so humbling himself, and aligning himself with the foolish laughingstocks of the world, Sidney deflects any accusations of pride in self, and flattery. Sidney's audience, however, cannot but be aware that this too might be a clever strategy, and if they are not careful, they might wish themselves a poet.

In conclusion, the reader of Sidney should not identify with the object of praise (to be a horseman or a poet) but to identify with the interpreter "who substitutes his own reading for the 'surface' meaning of the text."[40] In other words, the reader must be skeptical regarding absolute meaning of any utterance, including that of the logician who criticizes (or deconstructs) another's meaning, replacing one of his or her own. This replacement is motivated by self-love. The poet (or orator), on the other hand, simply suggests an idea to the reader, who then reconstructs an almost totally new idea which resembles, but is not identical to, the idea presented by the original poet. The reader, in so doing, makes a poem the "imaginative ground plot for a profitable invention."

Although Sidney might agree with Hobbes that names are assigned arbitrarily for convenience, he claims that poets realize that these assignments are a "truthful" counterfeit, an artful but honest lie and fiction. To paraphrase Sidney, philosophers and historians, in their vanity, imply that names, though arbitrarily imposed by the authority and fiat of either the Church, king, or communal consent, finally become identical to things through the ratiocinative process, through the arbitration of correct and consistent definitions, through a strict addition and subtraction of account. In the *Defence,* Sidney mocks this philosophers' Idol and articulates the usefulness of the honorable lie. To the charge that poets are liars, he answers paradoxically, but truthfully, that, of all writers, the poet is the least a liar, for as the poet nothing affirms, so he never lies.

> So as the other artists, and especially the historian, affirming many things, can, in the cloudy knowledge of mankind, hardly escapes from many lies. But the poet (as I said before) never affirmeth. The poet never maketh any circles about your imagination, to conjure you to believe for true what he writes. . . . And therefore, as in history, looking for truth, they may go away full fraught with falsehood, so in poesy, looking but for fiction, they shall use the narration but as an imaginative ground-plot of a profitable invention. (*Defence,* 52–53)

Similarly, Bacon challenges the deceitful philosopher who is content to live within the shadow of the truth. Judging from his essay "Of Truth," however, it is unclear whether anyone is capable of finding truth and presenting it to the world in the clear light of day.[41] In fact, in a later essay, "Of Simulation and Dissimulation," Bacon encourages the use of dissimulation by men of judgment and discretion. First, he claims that persons of small wit are habitual dissemblers; deprived of choice and variation in action, due to their feeble judgment, these dissemblers must take the safest and "wariest" path through life. However, men who are habitually open, frank and well-managed (like a properly schooled horse) may, under proper circumstances, "give the lie" and thereby save the appearance of truth. His formerly good and honorable dealings make the lie invisible. Under these conditions, a habit of closeness and secrecy is both political and moral. "And in this part it is good that a man's face give his tongue leave to speak" (Bacon, "On Simulation and Dissimulation," *Essays,* 77). In other words, it is more important that the people think a man honest than that he actually be honest.

"Tell a lie and find a troth" a Spanish proverb suggests, as if there were no way of discovery but by dissimulation, by feigning, resembling, and mimicking the truth. Of course, the danger in this habit, this course of artful simulation, to present a counterfeit for the genuine article, is to lose an indispensable instrument of action, namely trust and credibility. Bacon, therefore, presents a formula that safeguards against such a collapse in credibility for the artful politician: "The best composition and temperature is to have openness in fame and opinion, secrecy in habit, dissimulation in seasonable use, and a power to feign, if there be no remedy" (Bacon, "On Simulation and Dissimulation," *Essays*, 78). This is particularly useful in times of trouble, when "the politic and artificial nourishing and entertaining of hopes, and carrying men from hopes to hopes, is one of the best antidotes against the poison of discontentments" (Bacon, "On Seditions and Troubles," *Essays*, 106).

This kind of hermeneutical strategy is shared by Edmund Spenser, most notably in the *Faerie Queene,* where his use of ekphrastic devices foregrounds the imaging or visualization of a knowledge that can only be perspectival and partial. For Spenser, the speaking picture—which included fable, mirrors, portraits, paintings, and dreams—could succeed in beguiling and distracting the unwary knight/reader without some controlling and interpretive guide. As such, the same artistic representation, when crafted and interpreted by the guiding hand of the gentleman poet, is used to teach and instruct. In Book I, the villain Archimago, maker of images, creates a false lady to beguile the fantasy of the hero knight, Redcrosse. This new creature, artificially born without her due and full of the maker's guile, is taught to imitate the true lady Una, thus distracting the knight from his quest with false and abusive shows. Ironically, the poet, like Archimago, creates images as a vehicle of substitution, a scientific fiction that resembles a reality that is too large or indistinct to comprehend without such a concrete artifact. In Spenser's case, however, the poet sees himself not as a demonic villain, but as a gentleman who is imitating the function of the divine, as he (rather than his English Protestant society) understands it, using his poetic creation to teach, not to distract or seduce. This point is emphasized throughout the poem, notably in the episode of Book III which finds Britomart gazing into the glass of Venus. Here again, Spenser incorporates some kind of illusionistic device to caution the reader against seductive creations discovered in portrait, mirrors, and paintings.

In chapter 3, I shall question why such artistic devices (virtual incorporations) were used as "ways of knowing" in some cases, and dismissed as dangerous and misleading snares of the gaze in others. In this way, Bacon's own reform of learning, his attempt to construct and later dismantle the barriers between poetics, rhetoric, and scientific inquiry, can be thrown into relief, demonstrating how cultural and literary forces succeeded in shaping the philosophical and scientific debates of the time. I also investigate early seventeenth-century educational reform, focusing on the writings of Francis Bacon. For Bacon, the need to address—and, possibly, correct—the increasingly blurred distinctions between logic, rhetoric, and poetic would lead to the foundation of a new approach, a new advancement, in learning.

# 3
# Francis Bacon: The "Restauration"

> *What is truth?* said jesting Pilate, and would not stay
> for an answer.
> —Francis Bacon, "Of Truth," *The Essays*

IN CHAPTER 2, I SUGGESTED HOW THE ECLECTIC METHODological approach of Renaissance humanism successfully negated the distinctions and differences between poetics, rhetoric, and a dialectic that bound the verbal sciences to the theoretical and practical sciences of philosophy and history. Erasmus advocated a *copia verborum* that stimulated the senses through verbal variety. Sir Philip Sidney, following the example of Erasmus, used verbal ingenuity to test the reader's interpretive skill in detecting falsehood and duplicity, allowing rhetorical and psychological practices to perform logical functions. Logic as a distinct, scientific discipline yields to a dialectic indistinguishable from rhetoric. In an effort to combat the confusion of rhetoric and logic, Francis Bacon, in his *Instauratio Magna*, attempts a "restauration" of knowledge founded on new methods of discovery.

In October 1620, Francis Bacon published the first installment of his *Instauratio Magna,* the "Novum Organum," presenting a new instrument concerning the interpretation of nature.[1] In his position of authority, having been appointed Lord Chancellor two years earlier, Bacon launches an ambitious scheme of intellectual renewal, an instauration designed to restore to its perfect and original condition the commerce between the mind of man and the nature of things. Bacon's new epistemology, the experiential method, attacked the ancient art of eloquence whereby truth could be discovered and transmitted through the closed form of a book. By emphasizing things over words, Bacon criticizes the linguistic practices of both humanist and schoolman, repudiates current intellectual habits, and produces a method of scientific induction based on the nature of things themselves.

The seeds of Bacon's restauration of knowledge were sown in his boyhood, under the influence of his father's early attempts at educational reform. Sir Nicholas Bacon had been commissioned by Henry VIII to devise a new plan of university organization, preferably under lay administration. In this scheme, later adapted and refined by Sir Humphrey Gilbert, natural philosophers were to become professional researchers, seeking evidence drawn from nature and delivering their experiments in a plain language "without equivocations or enigmatical phrases."[2] Although his father's reform never came to fruition, Bacon did not cease in his aim to gain political authority through a similar administrative post, using a similar epistemological plan. Bacon used his program of scientific reform as an inducement to gain political position, soliciting the help of his uncle, William Cecil, Lord Burleigh.[3]

Bacon's political career began in earnest with his attachment to the Earl of Essex in 1588. Although Essex was unsuccessful in gaining a Crown post for Bacon, he did bring him to the Queen's notice by providing masques and entertainment, several of which were written by Bacon. When these government positions were continually denied him, Bacon threatened to relinquish his political ambitions for a contemplative life of study.[4] Upon the accession of James I to the throne of England in 1603, Bacon's hopes for political favor were reengaged. Although troubled by the execution of Essex, and by his own implication in that affair, Bacon was appointed King's Counsel in 1604, and Solicitor General in 1607.

Throughout this period of political frustration (1588–1604), Bacon continued to lead an active rather than a contemplative life, writing short treatises and entertainments, delivered in a courtly style, witty and ingenious, designed to bring his claims to the attention of the Queen. The first volume of *Essays, Colours of Good and Evil,* and *Meditationes Sacrae* were published in 1597. All three obviously were aimed at the courtier who, like Bacon, found the need to balance an active with a contemplative life. As such, in these works, Bacon employs old conceits to work his new ideas of moral, political, and philosophical reform, casting his aphoristic wit in an appealing and persuasive form.

In the politically fruitful years between 1603 and 1609, Bacon published only two works, *The Advancement of Learning* (1605) and *De Sapientia Veterum,* or *The Wisdom of the Ancients* (1609). Both address Bacon's plan for scientific and philosophical reform, and are therefore concerned more deeply with the discovery and transmission of knowledge than were his previously published works. In these two texts, however, Bacon still combined new conceits with old, devising a temporary compromise without alienating public opinion. As Farrington notes, "Bacon was publishing about as much as he thought his public could take."[5]

Bacon's unpublished works show a more violent, polemic attitude toward the traditional institutions of learning and the intellectual capacity of both academician and layman. In the early unpublished treatises, *Temporis Partus Masculus* (*The Masculine Birth of Time,* 1603), *Cogitata et Visa* (*Thoughts and Conclusions,* 1607), and *Redargutio Philosophiarum* (*The Refutation of Philosophies,* 1608), he formulates many of the arguments against traditional learning found in the later *Instauratio Magna.*[6] In the *Novum Organon,* Bacon writes that the entire fabric of human reason is badly put together, largely due to the vanity and wandering intellect of past philosophers. One group of philosophers, the dogmatic Academicians, overrate the store of knowledge, making the work short and easy, thus appealing to the vanity of popular opinion. The second group, the Skeptics, underrate their intellectual strength and continually complain about "the subtlety of nature, the hiding-places of truth, the obscurity of things, the entanglement of causes, and the weakness of the human mind."[7] Bacon proposes an alternative method of intellectual inquiry derived from the more ancient of the Greeks, the pre-Socratics, whose writings were lost, or buried, by the Platonic Academy. These philosophers take up with better judgment

> a position between these two extremes, —between the presumption of pronouncing on everything, and the despair of comprehending anything ... they did not the less follow up their object and engage with Nature; thinking that this very question, —viz. whether or no anything can be known, —was to be settled not by arguing, but by trying." (*NO,* p.256)

This path toward knowledge would be settled by hard thinking and a perpetual working and exercise of the mind, starting from simple, sensuous perceptions.

Similarly, in his earlier unpublished works, Bacon argues that Aristotle formulated a manual of madness, making natural philosophers the slaves of words. Plato turned scientists away from things, teaching them to turn their mind's eye inward, to grovel before their own blind and confused idols under the name of contemplative philosophy. Evidence, which should be drawn from things in a careful unmasking and sifting process, is instead clouded by a preordained scheme of interpretation. "Do you suppose, when all the approaches and entrances to men's minds are beset and blocked by the most obscure idols—idols deeply implanted and, as it were, burned in—that any clean and polished surface remains in the mirror of the mind on which the genuine natural light of things can fall" (*MBT,* 62). In keeping with his own political and administrative goals, Bacon argues that scientific reform must be linked to an active, worldly life, that it must be initiated by the experienced and learned man of affairs. In his role as guide, handing over the torch of science, Bacon asks for nothing less than to choose and polish the mental fabric of his followers. He demands as well that the general fabric of education, the conditions of learning, be repaired in the process:

> Do you really think it is easy to provide the favourable conditions required for the legitimate passing on of knowledge? The method must be mild and afford no occasion of error. It must have in it an inherent power of winning support and a vital principle which will stand up against the ravages of time, so that the tradition of science may mature and speak like some lively vigorous vine. Then also science must be such as to select her followers, who must be worthy to be adopted into her family. (*MBT,* 62)

Bacon's strategy is to attack the basic principles of Aristotelian philosophy as supreme examples of error and factionalism, leading to the corruption and degradation of both knowledge and the learning process. He blames Aristotelian intellectualism for obscuring the partial redemption of learning procured by the predecessors of Homer and Hesiod, the remote ancients who represented the face of nature, the stamp of God, in hieroglyphics, gestures, and fables—all of which have some similitude to the thing signified and function like emblems, icons, or pictures. Indeed, Bacon's restoration of learning does not completely disregard past practices. The present, he writes, is like a seer with two faces, "one looking towards the future, the other towards the past" (*MBT,* 68). The "sons of science" should find inspiration from the pre-Socratic philosophers, Heraclitus, Democritus, Pythagoras, Anaxagoras, and Empedocles, who, without noisy advertisement or professorial pomp, deduce truth from scientific analogy and experience, from a chaste, holy, and legal wedlock with *things themselves*. In this jointure of mind and nature, divine mercy acts as bridewoman (*MBT,* 72; *RP,* 111, 131). Unlike the superstitious popularizers, the Academicians and Sophists who study nature from a distance and construct the arts of disputation like ingenious perspective glasses, the true scientist should not appeal to the popular imagination, but instead should climb down from the tower and look objectively at things themselves (*RP,* 129).

As such, in his published and unpublished writings of the period 1603–9, Bacon continually experiments with the intersection of sensual experience and mimetic representation, exploring the heuristic and hermeneutical possibilities presented by figurative language. Most notable is his gradual allegiance to the tradition of allegorical exegesis. In the 1603 *Temporis Partus Masculus* (1603), Bacon rejects the usefulness of ancient myth and fable ("This hunting after guesses is a wearisome business and it would not be a proper thing for me, who am preparing things useful for the future of the human race, to bury myself in the study of ancient literature," *MBT,* 68). In the *Advancement of Learning,* of 1605, his attitude is more ambiguous: "But yet that all fables and fictions of the poets were but pleasure and not figures, I interpose no opinion" (*ADV,* 82). By 1609, with the publication of *De Sapientia Veterum,* he is firmly convinced that the veil or dense mist of fable facilitates the modern advancement of learning. Already in the *Cogitationes De Scientia Humana* of 1605, Bacon is disguising his own ideas and doctrines under the veil of allegory, refuting his earlier ideal to seek scientific truths from the light of nature, not from the darkness of antiquity (*MBT,* 69). The philosopher's task is to find the hidden meaning concealed in myths.

The question remains, however: What functional place did these fables hold in Bacon's restauration of philosophical and scientific learning? Spedding, Ellis, and Heath, the English editors of Bacon's work, make a clear distinction between the philosophical and the literary writing, classifying his allegorical use of fable with the latter. Paolo Rossi claims that the ideas on fable and other tropes of resemblance found in the *Advancement of Learning* and *De Sapientia Veterum* are conventional when compared to the views expressed in the unpublished works.[8] In the unpublished texts, Bacon eschews the use of traditional material and copious language in the progression of knowledge. According to Rossi, with the *Advancement* and *De Sapien-*

*tia,* Bacon gives in to public pressure, realizing that, to appeal to the common palate, the new must be mixed with the old. John C. Briggs, similarly, finds a pattern in Bacon's use of mythological figures and characters as a means of communication rather than discovery, claiming that when he turned his attention toward natural philosophy, he used his "biblical paradigms of experience and code to encompass the matter and forms of nature."[9] Bacon had earlier sent the unpublished works to Sir Thomas Bodley for his comments. In 1609, Bodley replied that, "although I am convinced as to the contents and subject of this admirable work, in no place of learning would you find a tribunal which would be able to acquit you of error."[10] Possibly because of this rebuff, Rossi argues, Bacon modified several of his views, especially on the use of figurative and imaginative language in the representation of scientific knowledge. Bacon became convinced that ancient myths were too alluring and seductive to leave without interpretation. By placing these fables under the guiding hand of the philosopher, Bacon diminished the potentially divisive nature of myths.

In the remainder of this chapter, I will reexamine and complicate these claims by demonstrating the pervasive importance of ancient myth in Bacon's general reform of science and philosophy.[11] Bacon is indeed fascinated with the allure of ancient fables, and is intent upon explaining their affective appeal through a systematic rationalization of the correspondence between *res* and *verba* (truth and figure). His tactic is to distinguish first between the discovery of knowledge and the use of knowledge. Rhetorical arts, he suggests, are related strictly to the latter, proving useful in the transmission of knowledge. In the *process of discovery,* verbal ornamentation is indeed a handicap, allowing the senses to be deceived through the Idols of the Mind. In the 1605 *Advancement of Learning,* expanded and later elaborated in *De Dignitate et Augmentis Scientiarum,* Bacon further divides the discovery of arts into learned experience (*experientia literata*) and the new organon (*interpretatio naturae*).[12] Learned experience is not, in actuality, an art or a part of philosophy but a kind of sagacity which Bacon, following the ancient fable, calls the Hunt of Pan. It was Pan who, by a happy accident as he was hunting, discovered the hiding place of Ceres. In the *Advancement,* Bacon claims that this is no real invention, but "a remembrance or suggestion, with an application" (*ADV,* 122). Yet, because learned experience takes place in a forest at large, Bacon includes it among the parts of invention, "so as it be perceived and discerned, that the scope and end of this invention is readiness and present use of our knowledge, and not addition or amplification thereof" (*ADV,* 122). This is an inferior method, whereby the scientist is led by a guiding hand, using some direction and order in experimentation.

In contrast, the light of truth itself is to be found in the new organon, in the interpretation of nature (*DAS,* Book V, ii, 505). This method proceeds by a true induction and utilizes the aphoristic method of transmission. Aphorisms, he explains later, are made out of the pith and heart of sciences, "for illustration and excursion are cut off; variety of examples is cut off; deduction and connexion are cut off: descriptions of practice are cut off; so there is nothing left to make the aphorisms of but some good quantity of observation. And therefore a man will not be equal to the writing in aphorisms, nor indeed will he think of doing so, unless he feels that he is amply and solidly furnished for the work" (*DAS,* VI, ii, 531). In fact, the first and most ancient seekers after truth stored knowledge in the form of aphorism.

In the section devoted to the preservation of knowledge in writing and the memory, (*DAS,* V, v, 519), Bacon implies that these aphorisms are created out of the mental images stored in the memory. Prenotion is a principal part of artificial memory, wherein the places are digested and prepared beforehand. This prenotion is an intuitive faculty that enables the scientist to complete the research, the "Hunt of Pan." This prenotion allows the mind to recognize order, and to fit the *word* with the *thing* itself. "But then we have a prenotion that the image must be one which has some conformity with the place; and this reminds the memory, and in some measure paves the way to the thing we seek" (*DAS,* V, v, 520). From here, the emblematic part of memory further reduces an intellectual conceit into a sensible image, "for an object of sense always strikes the memory more forcibly and is more easily impressed upon it than an object of the intellect" (*DAS,* V, v, 520). It seems, then, that inventions, including the invention of language arts in learned experience and the new organon, rely on the faculty of memory, and the peculiar operations that Bacon defines, rather vaguely, as prenotive and emblematic.

Using the example of the ancient hieroglyph, Bacon explains that it is not necessary that the cogitations of the mind be expressed by the medium of words. This type of communication resembles the gesturing of primitive people, and is distinct from any nominal system of signification. These "notes of things" are real characters, or symbols, representing neither letters nor words, but things and notions. Further, they function in two ways: *ex congruo* and *ad placitum*. Signs function by congruity where there is some similitude or congruity with the notion in the mind; these signs include hieroglyphics and gesture, the latter being a kind of transitory hieroglyphic which, like uttered words, fly away. The former act like written words, expressed in pictures; in their emblematic function, they have some similitude with the thing signified. As an example of an effective emblem, Bacon recounts the story of Periander who demonstrated to his messenger the nature of tyranny by lopping all the highest flowers in his garden. He implies that this type of emblematic and imaginative signification is the least ambiguous and is prone to error, simply because the picture conforms in an isomorphic way to the action represented. In contrast, notes which are *ad placitum* are agreed upon at pleasure, by contract and acceptation, by convention and usage. Bacon finds a deficiency in this portion of knowledge, in the imposition of names by contract. "Although some have been willing by curious inquiry, or rather by apt feigning, to have derived imposition of names from reason and intendment; a speculation elegant, and, by reason it searcheth into antiquity, reverent; but sparingly mixed with truth, and of small fruit" (*ADV,* 132).

Bacon further distinguishes between literary and philosophical grammar. The first concerns the pedagogical rules laid down for the instruction of a correct and pure habit of speech. In contrast, philosophical grammar "should diligently inquire, not the analogy of words with one another, but the analogy between words and things, or reason" (*DAS,* 523). Bacon asks that words, which are the images of things, be brought into some congruity with the things themselves. The philosophical grammarian should study and experience the actual use of the language, then draw from the several languages those things that are not only the most beautiful, but the most useful and the most precise reflection of the manners of the people. This composite language should be a model of some ideal speech, which expresses the meanings of the mind. The more readily the philosophical language resembles the thing or action (in the world), the more noble it will be. Furthermore, cogitations of the mind are not valid unless authorized by external actions perceptible to the senses.

With these statements, Bacon expresses his distrust of linguistic representation, his unease concerning any gap in scientific discourse between verbal character and perceptible, external reality. However, can we thereby infer that Bacon banishes completely the use of rhetorical figures, myths, and fables in the progression of scientific knowledge? Certainly, scholars frequently suggest that Bacon's method of scientific induction limits the active, similitude-generating faculties of the mind.[13] Bacon, however, contradicts this view on several different occasions. In Book VI, chapter ii of the *De Augmentis,* he identifies various methods of transmitting knowledge which explicitly endorse the use of similitudes and fables in the transmission of knowledge, and which implicitly encourage the reading and interpreting of ancient myth in the discovery of truth. Traditional material, he claims, should be used not only to teach popular opinions, but to inspire the progression of knowledge. In addition, he vacillates constantly between two epistemes, of the "innocent eye" and of the "knowing mind." In the latter, the memory uses a kind of intuitive faculty (prenotional) which guarantees a divine and authorized connection between experience and representation, between *thing* and *word.* Indeed, Bacon identifies the "digestion" of facts in the memory with prophecy and divine mercy, the latter acting as bridewoman in the wedlock of mind and nature.

In *Advancement* and *De Augmentis,* Bacon formulates his method of enigmatic insinuation in the progression of knowledge. He first makes a critical distinction between the magistral and the initiative (or probative) method; the first delivers the beginning of sciences, while the latter transmits the entire doctrine. Bacon argues that both methods are a part of logic, that they occupy a position immediately following the rules of invention. Although method, as a part of judgment, occupies a second place, it is, nevertheless, indispensable in the progression of knowledge.

> Neither is the method or the nature of the tradition material only to the use of knowledge, but likewise to the *progression of knowledge:* For since the labour and life of one man cannot attain to perfection of knowledge, the wisdom of the tradition is that which inspireth the felicity of continuance and proceeding. And therefore the most real diversity of method is of method referred to use, and method referred to progression: whereof the one may be termed magistral, and the other of probation. (*ADV,* 134; my italics)

Bacon claims that this last method (the initiative method of probation) has been corrupted by present practices. Currently, the deliverer of knowledge is less concerned that the form be examined, and more interested that it be believed. As such, the deliverer is content to satisfy the present satisfaction of the receiver, glory making the author conceal his weakness, and "sloth making the disciple lose his strength."

In contrast, Bacon proposes an initiative method which, by intimating and insinuating knowledge, encourages the intellectual capacity of the receiver, requiring that knowledge be examined. The initiative method (a term Bacon borrows from the sacred ceremonies) delivers knowledge to the sons of science and not to the crowd of learners:

> But knowledge that is delivered as a thread to be spun on, ought to be delivered and intimated, if it were possible, in the same method wherein it was invented: and so is it possible of knowledge induced . . . a man may revisit and descend unto the foundations of his knowledge and consent; and so transplant it into another, as it grew in his own mind. For it is in knowledges as it is in plants: if you mean to use the plant, it is no matter for the roots; but if you mean to remove it to grow, then it is more assured to rest upon roots than slips: so the delivery of knowledges (as it is now used) is as of fair bodies of trees without the roots; good for the carpenter, but not for the planter. But if you will have sciences grow, it is less matter for the shaft or body of the tree, so you look well to the taking up of the roots. (*ADV,* 135)

Julian Martin finds that, in his defense of an initiative method, Bacon ingeniously attempts "to harness firmly any voluntaries in natural knowledge and to secure the monarchical civil order from the corrosive, even "popular," political consequences of unmediated knowledge and unofficial knowledge-makers."[14]

Martin's claim in favor of Bacon's conservative views about political authority is difficult to square with the more conventional argument, articulated by Morris Croll, that Bacon in his style and thinking was rebelling, like his predecessors Lipsius and Montaigne, against the authority of the ancient authors, such as Cicero, who might have been more effectively used to uphold the monarchy, the aristocracy, and the status quo.[15] In his articles on the baroque style in prose and the history of "Attic," or anti-Ciceronian prose, Croll maintains that Bacon, Montaigne, Browne, Pascal, and Donne are "professed opponents of determined and rigorous philosophic attitudes" when it comes to writing and the thinking process.[16] Their intellectual process, as determined by their prose style, is characterized by spontaneity and improvisation, with no predetermined plan or a violation of any preconceived agenda, which can also be seen in Bacon's method of induced knowledge.

In that same essay on baroque prose, Croll claims that "the ambition of these writers was to conduct an experimental investigation of the moral realities of their time, and to achieve a style appropriate to the expression of their discoveries and of the mental effort by which they were conducted" (Croll, "Baroque Style," 231–32). According to Croll, early seventeenth-century prose, and thereby thought, is difficult to accept and understand (due to badly constructed sentences, loose and casual syntactic connections, digressions, overuse of semicolons and colons, frequent use of anacoluthon) because twentieth-century readers are trained "solely in the logical and grammatical aspects of language in interpreting the forms of style that prevailed before the eighteenth century" ("Baroque Style," 231). Before this Enlightenment triumph of grammatical over rhetorical ideas, the early seventeenth-century writers were using, and experimenting with, linguistic forms that expressed "the labor of minds seeking the truth," the purpose being "to portray, not a thought, but a mind thinking" ("Baroque Style," 208, 210). If, as Julian Martin writes, Bacon's purpose was to secure civil order from "unofficial knowledge-makers" and to protect against "the political consequences of unmediated knowledge," then Croll's theory of Bacon's endorsement of an improvisational mode of thinking and writing whereby the idea must be directly experienced and represented, should be reconsidered.

One way of reconciling Croll's views on Bacon's experimental prose style and Martin's notions on Bacon's conservation politics is to investigate how Bacon distinguishes between appropriate and inappropriate ways of delivering knowledge. Using Democritus as a guide, Bacon encourages the deliverer of knowledge to use frequent parables, tropes, and similitudes, "to make themselves conceived, and . . . to prove and demonstrate" (*ADV,* 137). Conceits that are seated in popular opinion need only be proved or disputed. In contrast, deliverers of ideas that are new and foreign must have some recourse to similitudes and translations so that important and illuminating paradoxes will not be rejected or passed over. "For it is a rule in the art of transmission, that all knowledge which is not agreeable to anticipations or presuppositions must seek assistance from similitudes and comparisons" (*DAS*, VI, ii, 532). It is interesting to note that, in the earlier *Advancement of Learning*, Bacon relates this kind of veiled, enigmatic discourse to divine writing. "So in divine learning, we see how frequent parables and tropes are: for it is a rule, that whatsoever science is not consonant to presuppositions, must pray in aid of similitudes" (*ADV*, 137). In the later *De Augmentis*, this rule is applied generally to the true method of the sciences, *de methodo sincera, sive ad filios scientiarum*.

In relation to this scientific method, Bacon further identifies both the exoteric and the acroamatic methods, both of which aim to separate the crowd of vulgar learners from the select group of scientists. The first (exoteric) method delivers knowledge in a more open and direct way, while the acroamatic method is more secret. According to Bacon, the acroamatic method was used with judgment and discretion by the ancient writers. In later times, however, it has been used for deceitful purposes to press forward counterfeit merchandise. "The intention of it however seems to be by obscurity of delivery to exclude the vulgar (that is the profane vulgar) from the secrets of knowledges, and to admit those only who have either received the interpretation of the enigmas through the hands of the teachers, or have wits of such sharpness and discernment as can pierce the veil" (*DAS*, VI, ii, 531). This statement is, in itself, problematic. Is Bacon ironically describing the cabbalistic method of discovery and discourse used by certain alchemists?[17] Or is he defining his own reform of the probative mode of delivery?

There is certainly a marked resemblance between the acroamatic method and the initiative method used by the sons of science, the learned men of affairs, to exclude the vulgar and profane. This group of initiates, as described in the *Advancement* and the *New Atlantis*, are indeed like a holy priesthood of disciples—inventing, judging, and delivering knowledge in a manner that allows for its progression. The only real difference lies in Bacon's identification of the exoteric method with an open insinuation of knowledge, whereby scientists deliver the truth of *things*, using the same method in which it was invented. As such, the most effective literary form in their induction is the aphorism—a concise, pithy and vigorous observation which, in its intellectual and affective impact, is both profound and sagacious. In limiting illustration, examples, excursions, deductions, and connections, the aphorism presents truth in a manner that invites further thought.[18] However, in his use of aphorism, in his listing of some good quantity of observation, Bacon does not banish completely the use of metaphor, tropes, schemes, parables, and fables. He is simply encouraging a judicious, prudent use of these verbal devices in the interest of weight, rather than *copie*.

Bacon claims that enigmas and figures which veil meaning act as safeguards against the superficial and easy anticipations of the profane and vulgar crowd of learners. There is an implicit endorsement here of presenting knowledge in a secretive, veiled manner, in an attempt to strengthen the sharpness, discernment, and resolve of initiated wits. Bacon praises the metaphoric style of Democritus in presenting and proving ideas that were new and strange to people's minds, laboring in his similitudes and metaphors for the sake of teaching. Myth and allegory, as well as other verbal figures, are used by Bacon to keep "truth" from becoming ordinary and prosaic, as a challenge for the readers to exert themselves in the pursuit of philosophical knowledge; he thereby makes education more meaningful and valuable. In *De Sapientia Veterum,* Bacon endorses the fables of Homer and Hesiod as models to be imitated, in contrast to the often-dry intellectualism found in Aristotle. By laboring in similitudes and metaphors, the deliverer of knowledge succeeds in progressing knowledge, polishing the mind and stimulating the senses to further inquiry.

It is the ambiguities of perception that require such a polishing, as well as such a close scrutiny of nature. In the *Redargutio Philosophiarum,* Bacon claims that nature presents to the gaze a certain picture of herself, "a cloudy semblance of a picture, in which all the minute differences of things . . . are blurred by distance" (*RP,* 129). Similarly, in the later *Novum Organon,* he calls nature a labyrinth:

> The universe to the eye of the understanding is framed like a labyrinth; presenting as it does on every side so many ambiguities of way, such deceitful resemblances of objects and signs, natures so irregular in their lines, and so knotted and entangled. (*NO,* 245)

The way out of this tangle is hard to practice but easy to explain: rejecting the mental operation that follows the act of sense, the mind must progress by stages of certainty, starting directly from the simple sensuous perception. "All depends on keeping the eye steadily fixed upon the facts of nature, and so receiving their images simply as they are" (*NO*, 253). In this process, Bacon withdraws his own intellect and allows the images and rays of natural objects to meet in a natural point; the excellency of wit has little to do in the matter (*NO*, 246). Bacon dwells purely and constantly among the facts of nature.

This procedure has a certain affinity with the hermeneutical process articulated by Erasmus, which I explored in the previous section. By constant rereading, the receiver of knowledge is absorbed by the text and becomes transparent.[19] "Meaning is guaranteed by the intuitive center of the self, which becomes identical to the voice at the center of the text. Of course, in the case of Erasmian and Augustinian hermeneutics, that textual voice belongs to Christ." With Bacon, the "text" ceases to be scriptural or, indeed, enclosed within the written form of a book, becoming instead the external world of *things*. As such, verbal representation must match, duplicate, and imitate the *things themselves*, "with the excellency of wit having little to do with the matter." Of course, like Erasmus, Bacon is aware that mental operations and sensual perceptions are prone to error, failure, and deceit. His experiential method is designed to correct these inherent flaws which resulted from the fall of man, and the subsequent fall of language. The senses, he writes, "supply the means of discovering their own errors" (*NO*, 250). Through the subtlety of experimentation, artificially devised, the errors of sense can be corrected. In this new experimental understanding of *mimesis*, meaning is not located solely within the referent; nor is it discovered and guaranteed by simple sense perception. Instead, the inquiring mind is taught to probe the ambiguities of vision.

Bacon constantly moves from a position whereby things are represented directly from sense experience to a view in which the mind mediates, digests, and interprets the data. He articulates his notion of *literate experience* in the *Redargutio Philosophiarum* (1608) as a path from sense to intellect, from the recording of experience to the formulation of direction and order in experimentation.[20] The scientist should resemble the bee, adopting a middle course between the ant (the empiric who gathers and consumes) and the spider (the rationalist who spins webs out of itself). The bee draws her "material from the flowers of the garden or the field," transforming it by a "faculty peculiar to herself." This peculiar faculty does not take experience unaltered into the memory, but "digest(s) and assimilate(s) it for storing in the understanding" (*RP,* 131). Later identified with prophecy, the digesting functions like the bridewoman, divine mercy, who stands between mind and nature. The scientific mind, then, has a prophetic obligation to represent the external world in an original and creative manner, one that encourages the continual progression of knowledge.

Later in the same treatise, Bacon proclaims that nothing requires greater courage than reflection on the enterprise, good fortune, and great exploits of our own age:

> The thunderbolt is inimitable, said the ancients. In defiance of them we have proclaimed it imitable, and that not wildly but like sober men, on the evidence of our new engines. Nay, we have succeeded in imitating the heaven, whose property it is to encircle the earth; for this we have done by our voyages. (*RP,* 131)

This new scientist must fashion experiments in a way that reflects and imitates the variety found in the external world. In his understanding of *mimesis,* Bacon does not depart radically from the definition provided by Aristotle in his *Rhetoric* and *Poetics*. In fact, Bacon's insinuative and acroamatic method, used by the sons of science to present their discoveries, shares with rhetorical and poetic theory a concern with metaphor in its mimetic and heuristic function.

*Mimesis* belongs traditionally to poetics—the art of composing lyric, tragic, and epic poems; according to Aristotle, it should be held separate from philosophy, and even dialectic. In contrast, the aim of rhetoric is persuasion and, through its link with dialectic, it is an art of second-order argumentation and proof. Imitation theory, however, does make an appearance in the *Rhetoric* via the mimetic and representative function of metaphor. Paul Ricoeur defines metaphor as a trope of resemblance, constituting displacement and substitution. Metaphoric language serves as a logic of discovery, whereby the writer masters metaphor to show new possibilities and new worlds. In this sense, metaphor is like a scientist's model used to trigger the scientific imagination. For Ricoeur, the scientific model is meant to resemble a reality that could not otherwise be envisaged because of its size and complexity. With this heuristic instrument, the scientific imagination must be taught to see "otherwise," to see within the similar the dissimilar.

In its referential dimension, metaphor has the unique ability to redescribe reality.[21] In this new scientific *mimesis,* the analogy must submit to reality, yet still admit a creative component in redescribing the essential nature of the world. Ricoeur identifies the tension that rests at the very heart of *mimesis,*

> It is only through a grave misinterpretation that the Aristotelian *mimesis* can be confused with imitation in the sense of copy. If *mimesis* involves an initial reference to reality, this reference signifies nothing other than the very rule of nature over all production. But the creative dimension is inseparable from this referential movement.... This paradox... holds together the closeness to human reality and the far-ranging flight of fable-making.[22]

Quintilian makes the point that imitation is linked to techniques of making; in grasping these techniques, the craftsman will learn not only to "invent" objects in nature that he has never seen, but he will also discover the complex operations of the interpretive mind.[23] This is true, due to the many choices and possibilities open to an artist in simply duplicating or copying the reality he sees and the reality that he knows. It is only in the process of making an object resemble, match or imitate something else that any artist becomes aware of not only the visible world, but the many options involved in interpreting that external world. It is not the "innocent eye" which makes these decisions, but the "knowing mind."

Bacon adapts many of these views in his reformulation of a new scientific *mimesis*. The senses must be corrected by the polished and active intellect, the knowing and experienced mind. As such, the writer of learned books, and especially the writer of a scientific treatise, should not focus the reader's attention solely on the seductive qualities of the language, namely their visual or aural impact. Rather, the stories should point to and thereby reveal, the essence or truth of things. In this process of interpretation, the scientific reader of fable must utilize the prenotional operation of the memory discussed earlier in this chapter, namely, that intuitive faculty which confirms whether the image has some conformity with the thing itself by striking the memory in an appropriate and affective way. In its relationship to the emblematic intention of memory, which reduces intellectual conceptions to sensible images, this prenotional operation guarantees the connection that constitutes meaning. Meaning is thus relocated in the exchange between the thing represented and the verbal picture, which, in turn, triggers an effective and instructive response in the memory.

In his division of prenotional and emblematic intention, Bacon reappropriates the semantics urged by Erasmus, while adding allegorical and philosophical dimensions. For Erasmus, meaning is created and understood using the affective faculty, whereby the reader is physically engaged with the surface structure of the text. "For since things are learnt only by the sounds we attach to them, a person who is not skilled in the force of language is, of necessity, short-sighted, deluded, and unbalanced in his judgement of things as well."[24] Yet, Bacon, in his concern that the fable not be enjoyed for the simple pleasure of its verbal and stylistic strategies, for the spectacle whose meaning is a mere illusion if divorced from the allegorical subject matter, moves beyond Erasmus in assigning a meaning to each fable that is univocal and singular, transparent rather than opaque. The reader's attention must not be diverted to the particularities of style (and the allure of superficial and sensual elements) and away from a transcendent, rational meaning.

In this, Bacon conforms to the conservative philosophical and allegorical hermeneutics found in Augustine's *De Magistro* and *De Doctrina Christiana*. In the Augustinian semiotic process, temporal words and signs are, in themselves, empty. Instead, the words signify a meaning that is not so much beyond them but contained within them, aiding the memory, reminding the conscious mind of the divine truth that lies buried in the human soul. Things of the external world, including words, are communicated to the soul via the bodily senses; "the next stage is the power of reason, to which the facts communicated by the bodily senses are submitted for judgment."[25] Even when no sound is uttered a statement is formulated in the mind. "And so in all speech we only remind, since memory, within which words inhere, by revolving them causes to come into the mind the very things of which the words are signs."[26]

Yet, while the eternal wisdom which is Christ can teach without words, Augustine still grants these temporal signs a unique status in mediating the path between man and God. Signs are not to be considered in themselves, as are things, but as images which "cause us to think of something beyond the impression the thing itself makes upon the senses," just as smoke causes us to think of fire, etc.[27] These natural signs are distinct from conventional signs established arbitrarily through usage and consent. In this group, Augustine includes the signs contained in the holy books, discourse that is often obscure and covered with a dense mist. Yet, the enigmatic and metaphorically obscure language of Scripture testifies to a deeper, more philosophical significance at the center of God's plan. By demanding that his people work and labor in their understanding of his divine will, God conquers their pride and arrogance while still allowing ease, delight, and pleasure. For example, in the *Canticles,* words are used metaphorically and figuratively, permitting illustration to be accompanied by sensual pleasure. The Holy Spirit, inspiring the writers of the text, ensures that future generations find repose through similitude, and pleasure through the difficulty of paradox.

Signs are either literal or figurative, and their obscurity is generally a result of man's ignorance. An understanding of languages and nature is essential to interpreting any figurative utterance. In this cognitive process, whereby metaphor is not only enjoyed but understood, the reader is invited to explore, to re-create and thus create a new set of similitudes. The metaphoricity of scriptural language tests the knowledge of readers, teases their ingenuity, and allows the exegete to experience God in all his variety. The world itself is a text made up of things, sensible and intelligible, that serve as signs. The principles of inference, definition, division, and eloquence were given so that philosophers might comprehend the truth of the things around them, both in the exterior, sensible world and the interior world of their own bodies and souls. This kind of knowledge leads inevitably back to the source—that is, to the creator, God.

Augustine acknowledges the positive importance of words and the Christian's duty to use them skillfully. He urges the exegete to be aware of the several rhetorical and grammatical forms of thought as a necessary solution to the ambiguities of the Scriptures. For a few students, the fruits of grace are so sweet, in and of themselves, that they have no need to be persuaded by pleasant speech; others must be soothed and pleased by the verbal ornaments of rhetoric. The orator must move the audience to perform a particular action, using the rhetorical triad of teaching, pleasing, and persuading. In the *genera dicendi,* or levels of appropriate style, the subdued style resembles a dialectical mode of proposition, proof, and conclusion. The teacher must introduce questions, produce exempla as a form of proof, and present some type of clear solution that eliminates difficulties but still allows other questions to be raised. As a general rule, verbal ornaments are not to be used, as the affections are not to be engaged. When using a moderate style, the speaker is free to use verbal devices to beseech his audience in believing a certain truth. Logical argument is not appropriate, rhythm and sound producing the pathos required. Verbal decoration is limited to the moderate style; that is what distinguishes it from the subdued and grand style. In contrast, this grand style is distinct for its force of spirit and emotion, whereby the choice of words is determined by the ardor of the heart and not by a careful choice of the intellect.

In his divisions between a low, middle, and high style, Augustine mirrors a conservative way of representing moral and social distinctions or conventions without necessarily excluding the popular dimension from consideration. Bacon, by following Augustine's distinctions on style and the use of allegory, does not separate himself as fully as Croll would suggest from a Ciceronian or Isocratean rhetoric geared to public consumption.[28] According to Croll, "the secrets of nature are made known only to attentive and collected minds, prepared by a long preliminary training in habits of exclusion and rejection; and even to them but partially, and in moments of rare illumination" ("'Attic prose'," 89). Croll continues by saying that the style appropriate to represent the acquisition of truth "owes its persuasive power to a vivid and acute portrayal of individual experience rather than to the histrionic and sensuous expression of general ideas" ("'Attic prose'," 89). Such a reading verifies Bacon's elitist views about who should be investigating the secrets of nature, and agrees with Julian Martin's argument that Bacon's scientific agenda ultimately protected a conservative status quo. One still needs to see, though, how Bacon delivered his philosophical views to a more popular audience, using several ideas on style and allegory adopted from Augustine.

# 4
# Bacon's Politics of Allegory

> Nor is that other point to be passed over, that the Sphinx was
> subdued by a lame man with club feet; for men generally
> proceed too fast and in too great a hurry to the solution of
> the Sphinx's riddles; whence it follows that the Sphinx has
> the better of them, and instead of obtaining the sovereignty
> by works and effects, they only distract and worry their minds
> with disputations.
> —Bacon, *Wisdom of the Ancients*

BACON ADOPTS SEVERAL OF THE AUGUSTINIAN IDEAS OF LANguage discussed in the previous chapter when arguing for the appropriate interpretive strategy employed in reading ancient fables for the delivery and discovery of knowledge. He makes clear the traditional link of *mimesis* and allegorical exegesis in the preface to his *De Sapientia Veterum*. The study of ancient myths, he writes, reveals hidden and involved meaning. Improbable tales, which exceed the natural ways of thinking and whose monstrous, extravagant nature could never be envisioned by man, cry out that a parable must be hidden below, waiting to be uncovered and excavated by the prudent and rational interpreter. As such, the tale should be read for its philosophical content, rather than its rhetorical embellishment. Using this rule, Bacon notes that the tales of various writers have common themes and subject matter. He deduces from this that Homer and Hesiod were reporting the myths told to them by the storytellers of their own time. This impresses Bacon more than if the stories of Homer had been original inventions of a particular age. The commonalities of subject matter convince Bacon that within the tales was something lofty and great, a meaning that transcends the particularities of any given time or culture:

> They must be regarded as neither being the inventions nor belonging to the age of the poets themselves, but as sacred relics and light airs breathing out of better times, that were caught from the traditions of more ancient nations and so received into the flutes and trumpets of the Greeks. (*DSV,* 823)

In other words, the reader of these myths must disregard the rhetorical embellishments used by the various poets—namely, the sound of the Grecian flutes and trumpets—and dig up the allegorical meaning nestled within.

> Parables have been used in two ways . . . to disguise and veil the meaning, and they serve also to clear and throw light upon it. . . . No force of wit can deprive us of that. Nor is there any man of ordinary learning that will object to the reception of it as a thing grave and sober, and free from all vanity; of prime use to the sciences, and sometimes indispensable. (*DSV,* 823)

The form itself tells the reader that something is hidden which must be brought to light. Even if the story has no purpose except to delight and amuse, it still serves as a puzzle to be deciphered; the pleasure comes in cracking the code. Indeed, fables, enigmas, parables, and similitudes are used as a method of making clear to the understanding the inventions and conclusions of human reason.

The observations made in the *De Augmentis* concerning the heuristic use of emblems and hieroglyphics by barbarous people, and the use of similitudes by the pre-Socratic writers, find a place here as well. Primitive people, Bacon writes, were

> then rude and impatient of all subtleties that did not address themselves to the sense. . . . For as hieroglyphics came before letters, so parables came before arguments. And even now if any one wish to let new light on any subject into men's minds, and that without offence or harshness, he must still go the same way and call in the aid of similitudes. (*DSV,* 824)

Those readers who cannot get beyond the pleasure afforded by the story are termed dull and leaden, easily pleased by superficial delights and illusions. This recalls Augustine's criticism of an unhallowed use of rhetoric for the purpose of gratifying human vanity. In this sophistic scenario, the philosophiser fails to move beyond the surface of the text. For Augustine, the delightful figures of rhetoric "were dream-substances, mock realities, far less true than the real things which we see with the sight of our eyes in the sky or on the earth."[1]

Like Augustine, Bacon breaks away from the strictly technical use of rhetoric to promote a more philosophical rhetoric that interprets visible forms to articulate a specific political and scientific agenda. This new "rhetorical" theory stressed the eloquence of realities (*res*) or truth at the expense of words (*verba*). Bacon argues that in a more primitive age, external *things* were indicated through the use of sensual phenomena, through the physical actions of men and women, through *muthos,* or plot. This displacement of idea for action constituted for the ancients the "meaning" of any story or fable. It is Bacon's task to get beyond the story, the actions of concrete and physical characters, to the abstract idea that is buried beneath, thus enabling a second displacement and exchange to occur. In this way, Bacon repairs and restores the original emblematic intention of the ancient stories. He anticipates the failure of signification that might take place within the specific trope of resemblance and effaces any tension between the figure and its referent.

As an interpreter, Bacon always seeks to pin down the meaning of the text, to limit uncertainty and demonstrate clarity and communication. He substitutes for the myth an explanation or interpretation that follows the dictates of human reason:

> Men of no experience in affairs nor any learning beyond a few commonplaces, have applied the sense of the parables to some generalities and vulgar observations, without attaining their true force, their genuine propriety, or their deeper reach. Here, on the other hand, it will be found that though the subjects be old, yet the matter is new; while leaving behind us the open and level parts we bend our way towards the nobler heights that rise beyond. (*DSV,* 824)

As a man of experience in affairs and learning far beyond a few commonplaces, Bacon presumes to say, he is well equipped to supply the sense of the parables.

Bacon then proceeds to decipher the ancient fables in a dogmatic and openly opportunistic way, promoting his own scientific and political agenda. He employs myth to articulate ideas regarding the superiority of materialistic naturalism (the myths of Pan and Cupid), the advantages of political realism (the myths of Metis, the Cyclops, Endymion, Narcissus, Perseus, Actaeon, and Diomedes), the necessity of scientific research (the myths of Orpheus, Atalanta, and

Prometheus), and the proprieties of ethical and psychological conduct (the myths of Cassandra, Nemesis, Memnon, and the Sirens). For example, the myth of Cassandra, or plainness of speech, demonstrates the futility of speaking at random, with no regard for the audience. The myth is a reproof to those writers who waste their prophetic gifts by giving advice and admonition without regard to the consequences.

Using the myth of Pan, Bacon notes the congruity of names with the things they represent. Pan's name and goat-like form declare the universal form of things, or nature. His horns, rising to meet the heavens, demonstrate how the entire frame of nature rises to a point like a pyramid. In figure, Pan is biform, "on account of the difference between the bodies of the upper and the lower world" (*DSV,* 830). Pan's accidental discovery of Ceres indicates the scientist's hunt for sagacious experiences. Finally, in his marriage with Echo, Pan performs the basic task of nature in a true philosophy, which, by some coincidence, reflects and matches the perfect and true induction of Bacon's *Instauratio Magna*.

> For that is in fact the true philosophy which echoes most faithfully the voice of the world itself, and is written as it were from the world's own dictation; being indeed nothing else than the image and reflexion of it, which it only repeats and echoes, but adds nothing of its own. (*DSV,* 832)

Bacon humbly asserts that the true philosophy found in this ancient fable merely reflects the echoing voice of the world itself. However, given the opportunistic use he makes of these fables—interpreting them in light of his own philosophical, scientific, and political theories—the modern reader may question this assumption. In the second part of this chapter, I question further Bacon's ideas of authority in the interpretation of classical fables, examining how these myths were used by Bacon to further his own political agenda, and asking why, in his "restauration" of knowledge, certain tales proved more useful than others.

In the first book of the *Advancement of Learning,* Bacon ostensibly condemns the humanist habit of winning and persuading public opinion through eloquence and rhetorical colors, in order to verify and convince that audience of a particular truth claim. These humanists, in their regard and love of words, he claims, began to neglect things, and in so doing were in danger of mistaking an image for reality. Yet, in so doing, Bacon does not necessarily advance the idea that things are superior to words—or, indeed, that reality can be reduced to naked things at all. Waswo and Terry Eagleton both imply that, for Bacon, a close analysis of things will naturally lead to the discovery and invention of the proper words, which will translate, in a naked and transparent way, the true nature or essence of that reality. Similarly, Martin Elsky sides with the view that Bacon participated in—indeed, initiated—that rupture in English linguistic thought which Michel Foucault describes as the classical episteme of identity, as opposed to the Renaissance episteme of resemblance:

> An idol language, then, such as one based on written hieroglyphs, would clearly and transparently represent the things it is instituted to signify. The idea that letters and words are hieroglyphs coextensive with creation and possess an ontological status of their own as things would make no sense to Bacon, who helped put the visual linguistic sign on a new footing.[2]

This view is reiterated by Sidney Warhaft and Margreta de Grazia, who argue that Bacon rejected the idea of occult resemblances or spiritual meaning encoded like signatures within the works of nature.[3] These scholars claim that, whereas Bacon sees a providential order in nature, as well as a divine plan imprinted there as signatures, he does not believe in magical analogies and resemblances. Rather, these signatures, properly read, show a logical, causal relationship among the things of the world. John Briggs and Julian Martin, in differing studies about Bacon's rhetoric of nature and his use of traditional models to uphold a conservative view of the state, argue that Bacon's use of figures is designed for a particular political and social purpose.

Of course, within their arguments there are obvious contradictions; indeed Elsky and Warhaft, as well as Paolo Rossi, acknowledge that in Bacon's philosophy, science and magic continually overlap; magical, hermetic, even Platonic influences, can be found throughout his work. Indeed, Rossi, in the tradition of the historian Walter Pagel, claims that the polarities between an occult and a scientific mentality have been falsely defined by historians of science, literature, and language.[4] In fact, no such fissure is present in the period; both attitudes are present simultaneously within the work of one writer or a group of several writers. In the work of Robert Fludd, for instance, we can find the influence of Agrippa, Paracelsus, the Cabala, hermeticism,

Rosicrucianism, and an allegorical treatment of the Bible. Elsky openly admits, as does William Sessions, that Bacon's linguistic thought should be seen within the framework of the linguistic theories of Alexander Top and Guillaume Du Bartas, who also influenced George Herbert's ideas of signs and their reference.[5] Top and De Bartas both assert that written signs act as a base and a foundation that signify the order of nature. Word and thing are thereby attached in an almost magical, providential accord whose authority is the law of divine imitation. The faculty of speech, the Word, attests to man's resemblance to God.

Without denying that Bacon's philosophy was radically new, in his own conception of it and in the subsequent use made of it by his intellectual followers and heirs, we can justify an investigation of how his approach to logic, as the study of the formal structures of thought, was still attached to the traditional arts of rhetoric. Bacon, along with his contemporaries, believes that Adam, when naming the animals, was able to relate accurately word to thing, thereby revealing the essence of the natural object. In *The Advancement of Learning*, he does indeed propose a restoration of Adam's tongue, discovering a finer and neater congruity between word and thing. However, Elsky writes: "for Bacon the restoration of the Adamic language has little to do with, say, Top's more traditional Cabalistic view of that language."[6] Bacon's point is that reality itself is not properly known, therefore the words abstracted from misunderstood phenomenon, and then applied to another misunderstood phenomenon, can only lead to mental confusion and disorder. According to Elsky, Bacon's concern here is a result of a mentality, an attitude toward the natural created world, unsupported by the view that all creation is based on analogy and resemblance. In this scenario, any application between word and thing will be a misapplication. Although Elsky makes an important point, here, Bacon himself was in a bit of a dilemma. Certainly, he does not give credence to the idea of a divine status between reality and speech, or to an ontological basis of resemblance in nature in terms of a microcosm or macrocosm. Bacon, however, cannot release himself from conceiving on the basis of comparison, analogy, and resemblance. This is the crux of the problem: Where and how are the comparisons being made?

According to Elsky, Bacon is unconcerned with any link between the physicality of writing and the concept these words signify. "In a sense, for Bacon language in general is like this system of ciphers; it parallels the world of things but has no contact with it except insofar as the univocal, arbitrarily assigned words refer to those things. One word is as good as another; each must simply be established by the consensus of its knowing users."[7] As such, Bacon is severing, in a new and radical way, the word from "the same social and ontological substratum as the thing it signifies" (179). Language, in other words, is only a copy or representation of the object, and is, therefore, a reflection of a reality that is wholly and forever distinct. The written word reflects the world not by any divine or natural correspondence, but by fiat, by the authorizing experiments of the inductive scientist. The link between word and thing will always be provisional and arbitrary, sustained by an artificial collusion between authorizing agents and knowing users. That agent, or group of agents, has successfully penetrated the secrets of both nature and the human intellect, clearing the enchanted glass of the human mind from all distortions and thus providing a transparent medium that will clearly reflect the nature of things themselves. Or, as Judith Anderson writes "for Bacon at least, words without material warrant have simply become unreal."[8]

The problem of subjectivity in this process is addressed by Timothy Reiss:

> The remaining ambiguities require a solution. Some kind of appeal to authority will provide it. Again, the difficulty is to conceal such an appeal. How to show simply that the logic of the excluded middle is founded in the true relation of concept and object, itself confirmed by the proofs of utility and practice, and that it is not simply the subject's fiat? The authority must be that of things themselves and of their seeds. The problem therefore is to displace the authority, to guide it toward its own disappearance (which was indeed one of the aims sought through the contractual installation of the liberal state). It is a matter, if you will, of concealing the limits of the new discourse.[9]

Yet, in his wisdom of the ancients, Bacon neither conceals nor displaces the authority of his text; rather, he replaces it within the methodological domain of ancient fable and allegory.

In her recent studies of the fables of power, Annabel Patterson claims that the fable, like the allegorical tale, has traditionally been used by the party in defeat, employed as a self-protected mode of communication given the sanction of an ancient form: "Those without power encode their commentary not to preclude understanding but to claim for their protest the sanction of an ancient form."[10]

Jonathan Goldberg likewise traces the etymology of allegory (*allos-agoreuein*) as meaning a speaking beside the point, cryptically and in private. It is the voice "present" to itself, not out in the open; the voice of the veil.[11] For Maureen Quilligan, allegory is a type of literary criticism which grows out of the particular need to posit a distance between the literal level and the allegorical significance, "specifically the defense of Homer against the moral attacks of the Socratic philosophers; wishing to renounce neither Homer nor philosophy, the Pergamean school read the epics to mean things other than they at first meant."[12] Allegorical interpretation is a type of defense when the literal meaning is too disturbing or too banal to be digested. In fact, the greater the separation between surface meaning and allegorical reference, the more clever is the allegorist for hiding the truth from vulgar eyes.

Paul de Man, in his article, "The Rhetoric of Temporality," gives a reinterpretation of the relationship between allegory and symbol as modes of language, and as a means toward self-discovery and self-negation.[13] The trend of Anglo-American criticism has been to subordinate the allegorical mode to the symbolic, following the literary aesthetic and taste which was fostered by post-eighteenth-century writers on nature. In this literary climate, where language was ideally seen, and used, as an instrument of synthesis and totality, the symbol performed a function quite distinct from that of allegory. The symbol, by supposedly linking experience and the representation of that experience, achieved a "total, single, and universal meaning" that transcended any particularities of place, time and contingent circumstance.[14] In contrast, in allegorical discourse, signs refer to a specific meaning whose potentiality and suggestive life is exhausted as soon as the sign has been decoded and interpreted. As de Man notes, this tendency toward an allegorical/symbolic dichotomy finds full expression in the age of Goethe. Such an antithesis was unimaginable to an earlier writer, such as Winckelmann, who continued to use the two terms interchangeably.

This antithesis survives well into the twentieth century and the literary aesthetics of Anglo-American criticism. Symbol, with its consequent stress on metaphor and imagery, is seen to be the fullest expression of poetic and linguistic achievement. In this scenario, allegory is often relegated to the realm of non-art as a dry, rational, dogmatic mode of reference whose meaning is disengaged from the linguistic form. Forever speaking beside the point, allegory is based on an arbitrary, often nonsensual decision of the mind. Again, de Man notes that by suggesting a disjunction between language and reality, "allegory designates primarily a distance in relation to its own origin, and renouncing the nostalgia and the desire to coincide, it establishes its language in the void of this temporal difference."[15] In other words, allegory continually reminds (and cautions) the reader against any full identification of language and reality, self or nonself. Or, in the case of sixteenth- and seventeenth-century writers, between the man-made artifice and the natural world. Allegory always unveils temporality by noting the gulf, the lack of resemblance, between the abstractions of verbal representation and the natural world.

De Man's primary aim is to reread the customary picture of early Romantic literature, finding in these writers an ambiguity regarding symbolic discourse as a totalizing experience, an unease concerning poetic symbol that links them to the Renaissance tradition of allegory, which posits a significant, insurmountable gap between verbal representation and the natural world this language refers to. Consequently, the self is constantly reminded of the gulf between itself and the natural world, between the creative act that brought forth that world, and the creative act that formulates a descriptive verbal impression. There is, indeed, little difference then between this early Romantic tendency and the lament of Augustine, that his words will never fully match his thoughts.

The poet's dilemma remains, forcing each writer to formulate a viable linguistic solution that will provisionally bridge that gap between words and reality, or things, and yet still voice the dilemma of temporality. Writing on language, Ernst Cassirer formulates a useful model to describe the development of language. The first original, or primitive stage, is the mimetic, whereby word and thing are nearly identical, onomatopoesis ensuring that verbal signifiers reproduce the sensual impression of natural objects. The epics of Homer and Hesiod, as well as the poems of Ossian, fall within this phase of linguistic development. Second is the analogical stage where words become detached from the things they refer to, while still retaining a resemblance to them. The rationalizing tendencies of the Greek academics can be associated with this movement. The final phase is the symbolic one, whereby words are arbitrarily assigned to abstract meanings. De Man, likewise, distinguishes three forms of language. Symbolic and mimetic representation are described as mystified forms that hide, or deny, the discontinuities which exist in the relationship of language to reality, self to nonself. The allegorist, with its counterpart, the ironist, demystifies this relationship, acknowledging the void of temporal difference in the very doubleness of their reflective vision.

That Bacon, in the seventeenth century, should turn to the allegorist, then, for guidance, and for a prior articulation of his own ideas, is particularly surprising.[16] However, as I argue in the first section of this chapter, Bacon relies heavily on a literate experience of nature (*experientia literata*), a hunt of Pan which necessarily precedes the other interpretation of nature (*interpretatio naturae*) or the new organon. This literary experience is linked to the philosophic grammar discussed at length later in Book VI of the *De Augmentis*. The literary kind of grammar or language is less useful in the discovery of knowledge, for, in this discipline, the grammarian merely investigates the analogy of words with one another, finding, interpreting, and analyzing the logical relationship between words in a given speech or discourse. According to Bacon, a superior kind of grammar is philosophical grammar which concerns itself with the analogy between words and things or reason, words being the footsteps of reason. Interestingly, Bacon asserts that this endeavor should stop short of that interpretation which belongs to logic. This statement usually is interpreted to mean that Bacon strictly separates the realms of grammar and logic, with logic being held superior and therefore in need of protection against the contaminating influence of words and rhetoric. Stephen Daniel, however, makes an interesting observation that Bacon is not protecting logic from grammar; rather, he is protecting his philosophic language from the corrupting influence of logical interpretation on the order of Aristotle and the Schoolmen:

> When the grammar of the poets is eliminated as a proper concern and tool of philosophy (as occurred in classical Greek philosophy and rational science), the discovery of useful works comes to an end. Socrates broke the chain which linked arts and sciences to one another and which permitted the mutually interactive extension of knowledge or "Circle of Learning" characteristic of the "universal *Sapience* and knowledge both of matter and words" taught by the poets and by the rhetorically sensitive early Greek thinkers.[17]

Bacon's early praise of the pre-Socratic thinkers, at the expense of the academicians Plato and Aristotle, illuminates his discussion here on the relationship of words to things and of rhetoric to philosophy.

The ancient mythmakers and the pre-Socratic philosophers, unlike the later schools of Plato and Aristotle, did not divorce the arts of poetry and rhetoric from the proper discovery of wisdom and knowledge—that is, science. Rhetoric was not left barren and ignoble by these thinkers who, in their epistemology, utilized the prerational techniques of the ancient poets. Things of the sense, articulated in poetry and myth, were tied more closely to things of the intellect, creating a "circle of knowledge," a procedure of discovery, that could regulate the epistemological presuppositions of subsequent ages, including Bacon's own. The literary mode of metaphoric and allegorical communication is not merely for display and ornamentation, but is crucial in preparing the mind for the investigation and discovery of the truths encoded in nature. Daniel is most insightful in his assertion that, for Bacon, nature must be treated as a poetic, metaphoric language rather than, as is usually thought, "a logical, mathematical, or rationally regulated language."[18]

The importance of metaphor and allegory as a way of training and polishing the mind for the later stage of interpretation, as a movement of thought that is necessary and appropriate for invention, is presented under the allegory, the hunt of Pan. In this kind of sagacity, the imagination is extended properly into several different domains; through a metaphorical or analogous transference of knowledge from one art to another, through a method of experimentation, proceeding by variation, production, translation, inversion, compulsion, application, or conjunction, the way is prepared for the invention of any axiom through the new organon. Particular arts and sciences, "disincorporated" from general knowledge by the academic philosophers, will act as helpmates in the invention of new knowledge.

> For I mean not that use which one science hath of another or ornament or help in practice, as the orator hath of knowledge of affections for moving, or as military science may have use of geometry for fortifications: but I mean it directly of that use by way of supply of light and information which the particulars and instances of one science do yield and present for the framing and correction of the axioms of another science in their very truth and notion.[19]

Such a translation (*translatio*) is urged by the orator Cicero who, complaining about the school of Socrates which divorced matter and words in the pursuit of knowledge, attempted, like Bacon, to fit rhetoric with philosophy in that great "Circle of Learning." The arts of rhetoric, then, are especially useful in handling and moving the minds and affections. Such handling is justified "insofar as nature itself exhibits or is seen as exhibiting a structure which is primarily linguistic rather than geometrical or logical" (Daniel, 222).

Nature must be experienced in a literary way, through literary explication, through the perception of analogy, inversion and variation of imagery, and through metaphoric enigmas or logical contradictions which require careful thought and reflection. Indeed, as allegorical stories alert us to the linguistic workings of nature, it is necessary to study the ancient mythmakers to justify the correspondence of prerational thought to modern epistemological modes of reference.

In *De Sapientia Veterum,* the wisdom of the ancients, Bacon combines tales with reason to produce a political and scientific allegory intended both to distance and familiarize the common assembly. Jon Whitman traces the history of the term *allegoria,* and notes the tension which exists between the two dominant meanings of the word.[20] In Greek, *allegoria* has two component parts—*allos,* meaning "other," and the verb *agoreuein,* meaning "to speak in the assembly," in the *agora*. This second component itself contains two connotations; to address an official political or legal assembly, and to speak in the open market, to a common or low audience. According to Jon Whitman, the first component inverts the sense of the second, "the resulting composite connot(ing) both that which was said in secret, and that which was unworthy of the crowd. These two connotations of the word allegory—guarded language and elite language—became explicit parts of allegorical theory and practice."[21] This inversion is especially relevant within the rhetorical tradition of allegorical interpretation. The stress here is on the tendency of allegorical compositions to "speak otherwise," to state one thing but mean another. The focus, both in terms of philosophy and exegesis, is to extract meaning from the allegorically written text.

In these cases, to call a text "allegorical" is, in itself, significant, marking a change in sensibility which Cassirer labels the analogic mode of thinking, distinct from the mimetic and symbolic modes. In this second stage, words become detached from their referent yet still resemble them in some way. James Calderwood notes the development of language articulated by Cassirer, from the mimetic to the symbolic phase, and demonstrates the Elizabethan dramatists' acknowledgment of a linguistic crisis.[22] In *Richard II,* for example, Shakespeare shows how word and thing, title and king, become disengaged. Whereas Richard possesses a meaningless, impotent title, Bolingbroke invades England "as a meaning in search of a name." Of course, Bolingbroke's "meaning" is not inherent in his person, but in his actions and intentions, his usefulness to England as a material force.

Alternately, Spenser uses myth to celebrate Elizabeth's reign, thereby justifying it in sociopolitical terms. Spenser interweaves the celebratory strain with the analytical.[23] While creating an image intended to dazzle and entertain, Spenser limits—indeed, conceals—real insight into the realities of the contemporary situation. By providing a logical, analytical structure that is like the discourse of law and justice, he rationalizes and justifies the mythological dazzle that would threaten and seduce the reader. In connection with this, Hulse argues that Bacon's demystification of myth is intended to warn the ruling elite of the dangers inherent in an ambiguous mythography that could be exploited by competing political cliques. Instead, Bacon mystifies "the operation and exercise of power, suggesting that command of state lies with those who command statecraft through their grasp of significant examples and their capacity to apply them to a present situation."[24]

Certainly, Bacon defends a political realism inspired by Machiavelli in various of his myths, such as the fables of the Cyclops, Metis, Endymion, Narcissus, Actaeon, Perseus, Achelous, Diomedes, the Styx, and Juno's suitors. By these fables, the king is warned to be wary of the people, and especially his ministers. Contempt for false flattery is displayed, although Bacon is ambiguous in his overall thought on flatterers. The fable of Juno's suitors warns that flattery may take various forms: an outward show of abjectness and degeneracy may constitute flattery, depending on the nature and character of those flattered. Although flattery is both necessary and inevitable, only a fool ignores the fact that flattery is no real sign of substance.

In this myth, and others pertaining to sociopolitical realities, Bacon does not mystify the operation and exercise of power, as Hulse maintains. Rather, Bacon uses fable not to conceal the real situation, in the philosophical or scientific realm or the sociopolitical one, but to reveal the necessity of further exploration, reflection, and insight. For Bacon, in a time before abstract nouns, things were indicated through the use of sensual phenomena, through the physical actions of men and women. Over the course of time, this primitive metaphor, this substitution of action for ideas was turned into a story. It is the scientist's task to move beyond the story, the sensual and active details, and find the structuring idea or philosophy behind the myth. Things of the senses are thereby linked to things of the intellect.

Elsky differs from Hulse by seeing in Bacon's interpretation of myth a demystification of the correspondence between language and reality, the exercise of power, which that language reflects and represents. Bacon negates this traditional role of allegory (as defined by de Man) which acknowledges the void of temporal difference between language and reality. Bacon's hieroglyphics, and his interpretation of myth, is "univocal rather than polysemous, transparent rather than opaque, and it has a clear and singular rather than an enigmatic relation to its meaning."[25]

For Bacon, a logical and arbitrarily defined relationship between words and things, an epistemology based on logical and analytical premises, would be an idol of the mind; in this scenario, reason becomes enthralled by the power of the human mind to fix such quick and easy correspondence between word and thing.[26] Instead, Bacon's epistemology of allegory and myth, his explanation and justification of the allegorist, does indeed echo that of de Man rather than the traditional definition of allegory. The allegorist, unlike the user of symbol and accurate mimetic representation (as articulated by Plato and his followers), openly acknowledges that there is a discontinuity between language and reality, a discontinuity that must be filled through literary hermeneutics and the heuristic models of the scientist. The fable is exciting and stimulating for the scientist, the person of insight who can probe, through literary explication and the *experientia literata,* the gaps in poetic language.

Because of the necessity of philosophical grammar in the interpretation of nature, Bacon limits the sensuality of the fables, warning his reader that to be charmed by the colorful exploits of the antique gods, goddesses, and heroes is to be enchanted by trumpets, pipes, and horns. Since that same myth can dazzle the vulgar crowd, stability must be arbitrarily set upon the myth; poetic language must be explained, made easy and accessible, and imaginative possibilities must be eliminated. This stability is maintained by reference to some outside reality, to the world "out there," in Ricoeur's phraseology. In his "poetics of things," Bacon may offer a solution to Ricoeur's engagement with the structuralists' literary grammar, or meta-language. Indeed, Bacon provides an excellent example of the shift from sentence to discourse; his "circle of learning" taken from Cicero might well be compared to Ricoeur's "discourse" (the arena of the text), that linguistic and critical space where the narrative is probed in terms of its generic and stylistic *dispositio.*

Bacon explains the scope and aim of his literary explication in the dedication to the Earl of Salisbury, Lord High Treasurer of England and Chancellor of the University of Cambridge. Bacon's aim is to give help in the difficulties of life and the secrets of science; if he is misunderstood the fault lies with the reader who has not obtained a deeper intellectual grasp of things but who has remained attached to the vulgar apprehension of surface details. The dedication is to the Chancellor as a scholar cognizant of both the art of politics and the matters of philosophy, which spring from the same fountain of learning. Politics, religion, and philosophy are all tied together in an interlocking network of relationships. And, within this grid, parables exist as a "kind of arc, in which the most precious portions of the sciences were deposited," a form of exposition that includes the primeval wisdom of antiquity which combines philosophy with grace, the eloquent husk acting as an ornament of life and the human soul (*DSV,* 821). There is certainly both a sacramental and a holy aspect to this idea of knowledge to be discovered and discerned only by an elite group of initiates. He seems to put forth explanation and method only as a last resort, to help the wider audience. Method is a univocal explanation used for the purpose of teaching. In contrast aphorisms intimate and insinuate ideas in the mind of the active seeker of truth in this hunt of Pan.

Lisa Jardine takes the view that, while Bacon may well have believed seriously in the philosophical truths hidden in myth, "there can be little doubt that in his moral and political interpretations he makes opportunistic use of the myths to communicate precepts in a persuasive form."[27] This observation gives us some insight into the way in which Bacon perceives the relationship between a representational type of aesthetic discourse and a more explanatory, philosophical, and discursive mode governed by reason. The latter epistemological stance works in tandem with the former, the only difference being that, in practice, Bacon's explanation can change with time and circumstance. It is ironic that this should be the case, that the philosophical explanation should be contingent upon circumstance, while the imaginative and figurative husk remains stable simply because of its enigmatic—indeed, ineffable—qualities.

This is certainly a perplexing dilemma, especially in the face of Bacon's stated claim, in the preface to *De Sapientia Veterum,* that the kernel of truth hidden within the myth or fable transcends any particular time, while the stylistic and aesthetic details, the sound of the flutes and trumpets of the Greeks, are particular to a given set of people. In Bacon's practice, in his decoding of the myth, we find that the opposite is true. He can interpret or read a myth in opposing ways, depending on the views he holds in different periods of his life, and in the practical repercussions of a particular reading.[28] This throws into question Elsky's idea that Bacon endorsed a method whereby knowledge should be conveyed clearly and unambiguously in terms of univocal and arbitrarily fixed definitions, agreed upon by a select group of scientific researchers.

The question, then, is: What does Bacon achieve in his interpretation of myth and fable? Certainly Bacon strives for and achieves a one-to-one correspondence between the details of the myth and the details of his own precepts regarding the aims of philosophy and science, as well as the advantages of a political realism on the order of Machiavelli. A close reading of several myths, however, reveals the impossibility of any representational claim regarding meaning. For example, in the myth of the Sphinx, Bacon carefully demonstrates how the Sphinx might well figure or represent science. This art of science is also, not absurdly, called a monster, in allusion to its many shapes and the immense variety of matter with which it deals. The beauty, eloquence, and facility of utterance found in a true science match that of the Sphinx, as do the wings, signifying the ability of scientific discoveries to fly abroad in an instant, and the claws, sharp and hooked, which reflect how penetrating are the axioms and arguments of science. In her enigmatic and riddling capacity, the Sphinx also resembles science which allows the understanding to be "free to wander and expatiate, and finds in the very uncertainty of conclusion and variety of choice a certain pleasure and delight" (*DSV,* 854). Bacon concludes by listing the many successful decoders of the Sphinx; both Oedipus and Augustus Caesar excelled in the art of politics by solving the many riddles concerning the nature of man presented by the Sphinx.

This particular essay, however, does not conclude with a neat, univocal and unambiguous meaning. Certainly, the myth ends with the

## CHAPTER 4: BACON'S POLITICS OF ALLEGORY 111

Sphinx on the back of an ass, subdued by the clubfooted man who made himself King of Thebes. Yet, Bacon conveniently neglects to add that the ultimate fate of Oedipus was far from victorious or triumphant. Also, Bacon ends the interpretation not with the triumph of man defeating the Sphinx; rather, the Sphinx ends victorious.

> Nor is that other point to be passed over, that the Sphinx was subdued by a lame man with club feet; for men generally proceed too fast and in too great a hurry to the solution of the Sphinx's riddles; whence it follows that the Sphinx has the better of them, and instead of obtaining the sovereignty by works and effects, they only distract and worry their minds with disputations. (*DSV*, 854)

Rather than end on a note of closure, Bacon opens up the discussion to future possibilities, of both error and insight.

It is more accurate to say that Bacon in the myths presents an emblematic portrait for the reader to imagine and then study. The details of this picture trigger related, associative thoughts in the mind of the viewer, thoughts derived from an awareness of the workings of nature, and the various incidents found in history books, poetry, and philosophical discourses. In the essay on Nemesis, or the vicissitude of things, Bacon shows this goddess in all her finery, with wings and a crown, an ashen spear, a phial with Ethiops in it, and sitting on a stag. He then translates these physical objects into another philosophical idea, namely the vicissitude of things, by noting how each object variously functions in another art or discipline. The wings, for example, translate into the sudden and unforeseen revolutions of things, an idea derived from the records of history. Bacon recalls that Cicero warned Decimus Brutus to beware of Octavius Caesar's bad faith and evil mind. The physical wings, then, are yoked to the idea of flight and change, which is further yoked to the betrayal and bad faith found in the account of Brutus and Caesar.[29]

This metaphorical linkage is demanded from the virtuoso scientist who must decipher language and nature to find a special meaning other than the literal one. Allegory, in addition to its literal meaning, possesses another metaphorical meaning, "wherein resides its capacity to be true as well as to provide the twist of insight we derive from some good metaphors."[30] The scientist must have the capacity to see and contemplate similarities which are vividly set before the eyes, in language and in nature.[31] As such, to ponder the details of myth and provide a literal explanation of it requires rigorous imaginative and mental training on the part of the exegete. Informative and didactic explanation does not take the place of imagination and feeling. On the contrary; imagination and feeling are vitally important in determining the meaning and operation of myth. Imagination and feeling complete and fulfill the cognitive penetration of both language and nature, to find the unity which the myth, on its own, encompasses.[32]

In conclusion, Bacon does indeed advocate a metaphorical sensibility in the decoding of nature, positing a view that the creative insight of the poet or mythmaker is a more vital tool in the discovery of knowledge than the logical and axiomatic reasoning of the mathematician. This argument does not, of course, address the issue raised by Jardine and others, namely that in his moral and political interpretations, Bacon opportunistically presents his own precepts and (conservative) ideology on power and government authority. Whether Bacon formulated his political and ethical views by objectively probing the books of nature and the ancients, or whether he imposed the traditional views of his culture onto these readings, remains an unanswered question, one that is beyond the scope of this study. What *is* apparent, however, is the provisional nature of these interpretations. Given Bacon's views on the delightful pleasures of skeptical uncertainty (which allows the intellect and the imagination to wander playfully and fruitfully over a variety of riddles and possible solutions), one can argue that the essays written and printed in the *De Sapientia Veterum* are meant to trigger or stimulate further ideas and associations. The printed work merely testifies to a metaphorical play of mind moving over a particularly vital and vigorous language, that of myth and allegorical interpretation. The hunt of Pan is never completed; the Sphinx does not remain on the back of the ass, subdued by the clubfooted man. The Sphinx keeps her final riddle. Like the authors of the *Arcadia* and the *Faerie Queene,* Bacon defers final meaning—asking the scientist, if not the general reader, to continually reenter the labyrinth of complexities. Although knowledge, once understood and published to the world, becomes dull and ordinary, the lame man, the scientist, must never be fooled into thinking that the Sphinx has been silenced.

# 5
# Wise Men's Counters: Visual and Verbal Knowledge in Hobbes and Boyle

> Apposite comparison not only give Light, but Strength, to the passages they belong to, since they are not always bare Pictures and Resemblances, but a kind of Argument; being oftentimes, if I may so call them, Analogous Instances, which do declare the Nature, or way of operating, of the Thing they relate to. . . .
> —Robert Boyle, *The Christian Virtuoso*

ACCORDING TO ANTONIO PÉREZ-RAMOS, "NO SCIENCE CAN EXist without a specific ethos and Francis Bacon was the first explicit and articulate theoretician of an ethics of science in the form of a message embodying values, visions, hopes, and rational expectations."[1] What is unusual and remarkable about Bacon's legacy is the relative lack of interest displayed by subsequent philosophers regarding the content of his work and, instead, the lengths to which his followers went to validate their own scientific endeavors as disciples of Bacon.[2] The Royal Society, founded in 1660, admitted members who were influenced by a range of natural philosophers and religious thinkers. On November 28, 1660, in the rooms of Lawrence Rooke, professor of geometry at Gresham College, following a lecture by Christopher Wren, professor of astronomy, a group of experimenters, thinkers, and writers which included Boyle, Petty, Seth Ward, Thomas Willis, Richard Lower and Robert Hooke, collected and formally constituted themselves into an academy for learning. The ethos of the institution cohered around what was determined as the founding methodology of Bacon and so-called Baconian science.[3]

In the manner of Bacon, these writers openly encouraged a plain style based on rational principles and objective knowledge perceptible to the senses. Thomas Sprat, spokesman for the Royal Society, separated the knowledge of nature from the colors of rhetoric, the devices of fancy, and the delightful deceit of fables, relegating the imagination to the realm of poetry, which was governed not by verity but by verisimilitude. Of course, this agenda, which was not practiced with any degree of consistency, served as an epistemological foundation for a broader political scheme. The appeal to reason and plain style was itself a rhetorical strategy used by Royal Society scientists, and defenders of the restored Church of England, to distinguish themselves from "enthusiastic" non-Conformists who abused the rational basis of language. Even within this group of writers, however, there were discontinuities of thought regarding *words* and *things*. In fact, the Baconian plain style based on *things* continued to be rich in metaphor and analogy. These writers appeal to the plain style in order to persuade their readers of the validity of their claims for truth.

Focusing on Bacon's inconsistent policy regarding metaphorical language and the proper correspondence between *words* and *things*, I will reenact and complicate the various critical and interpretive systems that were fighting for sovereignty in the period. Writers such as Hobbes, Ward, and Wilkins follow one aspect of Bacon's thought, which denies the importance of metaphor, establishes the arbitrariness of language and the supremacy of *things*, and urges the binary identification of the sign and the signified. These writers, however, appeal to a plain, naked style in an attempt to persuade the readership of the validity of their truth claims. Like the latitudinarian members of the restored Church of England, the natural philosophers of the Royal Society were anxious to remove themselves from the charge of mysticism and "enthusiasm" which, during this time, had been leveled against the non-conformists. Ironically, these proponents of reason, rationality, and plain style (Joseph Glanville, for example) continued to dabble in the occult, experimenting in alchemy. Furthermore, in the realm of language study, advocates of the plain style (for example, Robert Boyle and John Webster) based on rational principles, were pursuing an alternative thread in Bacon, founded on classical notions of *mimesis* and an Augustinean sign system which saw an inherent and divine, rather than arbitrary and contrived, resemblance between *res* and *verba,* words and things. My aim is to question these discontinuities in thinking and to ask whether certain writers fashioned various and contradictory language schemes in order to accommodate their own political and interpretive systems of thought.[4]

I will be searching for such ideas in the work of Hobbes, Boyle, Ward, Wilkins, Webster, Sprat, and other writers of natural philosophy of the period. I will also determine the link between the Royal Society writers and the restored Church of England. The latitudinarians of the restored Anglican Church appealed to the plain style, as did members of the Royal Society. How were the two groups linked, in terms of their political and social epistemology? Numerous members of the Royal Society were committed members of the Church of England, and they too feared a return of religious and linguistic "enthusiasm" and mysticism. This is especially true of Robert Boyle. My aim is to set these writers in some kind of political and social context and show how their experiments in science, as well as their experiments in the writing of prose suited a particular political and social outlook or situation. The Royal Society had soon cultivated enemies among traditionalists at Oxford and Cambridge who continued to stress the epistemological assumptions of conventional Aristotelianism. Brian Vickers notes that "the real issues . . . are, first the general dispute between the arts and the emergent sciences over which of them were truly 'useful' to society."[5] The traditional disciplines of philosophy, history, and rhetoric, the arts of the vita activa, were always thought to constitute a beneficent public life. After 1660, the emerging sciences, as articulated in the writings of Bacon, Hooke, Boyle, Sprat, and others, began to offer a challenge to these traditional arts, assuming a privileged position for the practical, utilitarian, and technological. All these endeavors of the new science were intended to promote progress and benefit human life. In his *History of the Royal Society* (1667), Sprat writes:

> I shall here present to the World an Account of the *First Institution of the Royal Society* and of the *Progress* which they have already made: in hope that this Learned and Inquisitive Age will either think their Indeavours worthy of its *Assistance*, or else will be thereby provok'd to attempt some *greater Enterprise* (if any such can be found out) for the Benefit of Humane life by the Advancement of *Real Knowledge.*[6]

One should note Sprat's emphasis on progress, assistance, greater enterprise, and real knowledge, ingredients designed to appeal to a public, and a monarch, who were interested in the broader world of increasing colonial expansion where practical knowledge of new things would be more useful than a continued reliance upon centuries-old authors and texts.[7]

A quick perusal of Sprat's list of topics also demonstrates the modernity of the enterprise:

> Part 1, Sect. XVI. *Modern Experimenters* ("The *Third* sort of *new Philosophers* have been those who have not onely disagreed from the *Antients* but have also propos'd to themselves the right course to slow and sure *Experimenting*"); Part 2, Sect. V. *A Model Of Their Whole Design* ("Their purpose is, in short, to make faithful *Records* of all the Works of *Nature* or *Art,* which can come within their reach"); Sect. VII. *It Consists Chiefly of Gentlemen* ("But though the *Society* entertains very many men of *particular Professions* yet the farr greater Number are *Gentlemen,* free and unconfin'd"); Sect. VIII. *A Defence of the Largeness of Their Number* ("All places and corners are now busie and warm about this Work . . . from the Shops of *Mechanicks,* from the Voyages of *Merchants,* from the Ploughs of *Husbandmen,* from the Sport, the Fishponds, the Parks, the Gardens of *Gentlemen;* the doubt therefore will onely touch *future Ages*"); Sect. XII. *Their Method of Inquiry* ("In their *Method* of *Inquiring* I will observe how they have behav'd themselves in things that might be brought within their *own Touch and Sight;* and how in those which are so remote and hard to be come by, that about them they were forc'd to trust *the reports of others*"); Sect XVII. *Their Judging of the Matter of Fact* ("Those to whom the conduct of the *Experiment* is committed, being dismiss'd with these advantages, do (as it were) carry the eyes and the imaginations of the whole company into the *Laboratory* with them"); and Sect. XXI. *Their Way of Registering* ("The *Society* has reduc'd its principal observations into one *commonstock,* and laid them up in publique *Registers,* to be nakedly transmitted to the next Generation of Men, and so from them to their Successors. And as their purpose was to heap up a mixt Mass of *Experiments,* without digesting them into any perfect model, so to this end they confin'd themselves to no order of subjects, and whatever they have recorded they have done it not as compleat Schemes of opinions but as bare unfinish'd Histories").

This list shows how the Royal Society saw themselves as initiating a new way of constructing a common stock of knowledge, practical, democratic, based on sensual experience, and eschewing any fixed metaphysical system, design, or pattern. All this coincides neatly with the ethos of Bacon, even to the extent that Sprat wishes for things to coincide evenly with words ("They have been most rigorous in putting in execution the only Remedy that can be found for this *extravagance:* and that has been a constant Resolution to reject all the amplifications, digressions, and swellings of style, to return back to the primitive purity and shortness when men deliver'd so many *things* almost in an equal number of *words*" 172).

The assumption made by Timothy Reiss, Richard Waswo, and others that Bacon's entire program is designed to make the language of the mind conform exactly and isomorphically to things is highly problematic. Whereas this kind of literal mentality is largely absent in Bacon, it is not without its advocates in the period. Two decades after the death of Bacon, another influential thinker and writer produced a text which again calls for a radical reevaluation of the way in which, in this case, political reality is conceptualized, articulated, and eventually instituted. Sixteen years before Sprat's *History of the Royal Society*, Thomas Hobbes published in 1651 *Leviathan, or The Matter, Forme, & Power of a Commonwealth, Ecclesiastical and Civill*, in order to reform the institutions of power which effectively govern a state, or civitas.[8] In so doing, Hobbes claims to be simply observing the workings of nature, mimicked by art, creating an artificial animal (the leviathan) called the commonwealth or state. Fittingly, Hobbes was a vocal critic of the philosophers who comprised the Royal Society, acting as the chief opponent to Boyle's experiments with, among other things, the air-pump, "mobilizing powerful arguments why the experimental programme could not produce the sort of knowledge Boyle recommended."[9] The real issue in this debate, though, proved to be the issue of language in the representation of nature, with Hobbes continually noting how cultural and political metaphors frame and mediate any and all observations or witnessing of nature. In *Leviathan*, Hobbes, using an artful reworking of the mimetic impulse, demonstrates playfully and in a highly metaphorical and allegorical temper how nature, the representative art by which God both created and now governs the world, is imitated by man in order to make (*techne*) an artificial object. In this process of making, man imitates himself, that is, the workings of his own physiological, emotional, and mental nature, to reproduce an artificial man that is the ecclesiastical and civil state, a body that excels the natural in strength and stature.

In his understanding of *techne*, Hobbes recalls Quintilian who claimed that by grasping the techniques of making, the craftsman will learn not only how to "invent" objects found in nature, but also how to discover the operations of the interpretive mind which can invent objects not present in the observable world. Hobbes continues by claiming that the head of this government, the sovereignty, is an artificial soul, giving life and motion; the joints are the judicial and executive magistrates; the artificial nerves are the state's notion of reward and punishment; the muscles are the rich and wealthy; the memory is provided by counsellors, while reason and will are representative of equity and the law. Concord is health in the body, sedition sickness, and civil war is death. The parable of the belly is thus reenacted by Hobbes to produce a total and artificial body figured on the natural workings of man's physical being. His treatise, then, is an analysis of the workings of nature applied to the political workings of the commonwealth, provisionally to recommend, like the good physician, the optimal conditions of good health, stability, and prosperity.

*Leviathan,* therefore, is conceived in much the same manner as an allegorical tale, or fable, one thing presented in terms of another. Usually, allegory grows out of the particular need to posit a distance between literal level and allegorical significance, serving as a type of defense when the literal meaning is too disturbing to be digested.[10] In fact, the greater the separation between surface meaning and allegorical reference, the more clever is the allegorist, for hiding the truth from vulgar eyes. In the case of Hobbes, however, the allegory is meant to clarify rather than obscure. The allegory does not create a distance between the natural object (the nature of man) and the artificial one (the commonwealth), but instead creates a situation whereby the two are held in parallel, serving as points of reference and identity. Sovereignty is not like the soul; rather, it is an artificial soul. The effect of this is to create two objects within the mind of the reader—one the natural body, the other the artificial state. By repeating and attaching the word *artificial* to the elements of the commonwealth, Hobbes forbids any misunderstanding to occur in the mind and understanding of the reader.

This mode of representation is in keeping with the general ethic of language as representation articulated throughout *Leviathan*. The text is divided into four parts. The first part is an analysis of the matter and the artificer, that is to say, man. The second part treats of how and by what covenants the commonwealth is made, the nature of rights and just power or authority of the sovereign, and what preserves and what dissolves the commonwealth. The third part concerns the Christian commonwealth, and the fourth and final section discusses the kingdom of darkness. For the purpose of this discussion, I will limit myself to the first section—the nature of man, as subject and as artificer. It should be noted that here Hobbes is not treating the commonwealth as such; rather he is analyzing the workings and the recurring failings of man, in the hope of discovering the possible defects and cures of the commonwealth. The purpose of the introduction, where he presents his case as an allegory, is to prepare the reader for the subsequent comparison that will be made later in the treatise, when he moves from a discussion of man, the maker and artificer, to the artificial object, the state. In such a way, the reader is to keep two images fixed in the memory, and will have cause later to see the resemblance between the two artifacts.

It is in the second chapter that Hobbes clearly articulates his understanding of the imagination, as distinct from the fancy. He begins by asserting that a body, once in motion, moves eternally unless some hindering force impedes that motion. When the wind ceases to blow, for instance, the rolling of waves continues for a time thereafter. Just so with imagination. An object presents itself to the senses and is impressed on the imagination, which, in turn, retains that image well after the object is removed from sight. Hobbes calls the state when that image fades and decays, "memory":

> This *decaying* sense, when wee would express the thing itself, (I mean *fancy* it selfe,) wee call *Imagination*, as I said before: But when we would express the *decay*, and signifie that the Sense is fading, old, and past, it is called *Memory*. So that *Imagination* and *Memory*, are but one thing, which for divers considerations hath divers names.

All sorts of images sit decaying in the storehouse of memory, which, Hobbes suggests, is a cold, dead place. (This, of course, brings to mind Sprat's idea that experimenters will add to the common stock of ideas held in the mind. What Sprat neglects to consider, and which Hobbes highlights, is the range of images already in the mind, which the experimenters, no matter how rigorous their method, only with difficulty will banish from their memories.) In the memory, Hobbes claims, the images of a horse and a man can be compounded to produce a Centaur. Similarly, a man can imagine himself to be a Hercules or an Alexander, if he in fact has taken to read too much of the Roman poets and historians. Such a compounded imagination is but a "Fiction of the Mind." Here, Hobbes is addressing the mimetic habit of sympathetic identification; the rhetorical strategies of the Roman poets, which are to present stories in vivid and dramatic detail, cause a fictional identification to occur in the viewer, the audience, the witness. This kind of mimetic identification can only be imaginary, fanciful, and fruitless, like the imaginations that come with sleep. Still, Hobbes' text could not be constructed without such use of the imagination.

The imaginations of those who sleep are called dreams, and are the result of objects or experiences that have come some time before, in the waking hours, either totally or partially through the senses. The stuff of dreams, then, caused by some distemper in the inward parts of the body, should be dismissed as such, as the result of a bad conscience, or, more readily, an eating disorder. One should certainly not reason by way of dreams. By the same token, the waking imagination functions like a waking dream; the images stored therein, and in the memory, are only an inexact impression of material objects. On the surface, Hobbes does not seriously consider the correspondence between material objects and the impression they leave generally in the imagination, either waking or sleeping. He openly claims that the connection between dreams and reality is so inexact that any cognition resulting from dreams should simply be dismissed as another bodily distemper. Again, however, the comparison of the state to a leviathan could only result from such an identification of mental image and sensory perception. Like Milton, Hobbes dismisses dreams as the workings of idle fancy, yet restores their validity in the quest for knowledge about the workings of nature and human desire.

Hobbes distinguishes understanding from fancy by further defining and refining his notion of imagination. Words press themselves upon the imaginative faculty; that is, voluntary signs, as words, are stored as images in the memory. When the images are put in a sequence, and their "contexture," their weaving together, properly understood in terms of negative and positive, then conceptualization can occur.

> The imagination that is raysed in man (or any other creature indued with the faculty of imagining) by words, or other voluntary signes, is that we generally call Understanding; and is common to Man and Beast. . . . That Understanding which is peculiar to Man, is the Understanding not onely his will; but his conceptions and thoughts, by the sequell and contexture of the names of things into Affirmations, Negations, and other formes of Speech. (*Leviathan,* 18)

This train of thought, or mental discourse, is of two kinds: unguided and without design, or regulated by some desire and design. The first kind of discourse is, of course, to be discouraged, just as fanciful imaginations with no reference to an external object must be prohibited. In such a stream of consciousness, any type of argument can occur, and any type of resolution adopted with little or no regard for logical truth and justice, either artificial or natural. As an example, Hobbes introduces a kind of reasoning that is possible in the present time of civil war—namely, the pertinence of asking the value of a Roman penny. The coherence of the logic is as follows: the king's (Charles I) ransom is related to the ransom of Christ, which is then related to the matter of thirty pence (pieces of gold) which sold Christ to his enemies. This price of treason is then brought into consideration when speculating on the price of the king's betrayal. For Hobbes, this kind of allegorical speculation and syllogistic reasoning, where one event is played out in the course of another subsequent event, is an abuse of the imaginative faculty whereby two unrelated occurrences are held parallel in an effort to obtain a satisfying solution to a present problem. Nevertheless, he makes the valid point that the human mind is prone to find resemblances between present concerns and past situations in books and stories. If, as Vickers suggests, the real issue lies in the "general dispute between the arts and the emergent sciences over which of them were truly 'useful' to society" (Vickers, 227), Hobbes offers a clear challenge to the epistemological assumptions found in Sprat, which constituted the great enterprise of the Royal Society in the advancement of real knowledge.

The second mode of mental discourse is to be preferred, namely a train of thought regulated by design. In this model, one reasons in a constant and consistent fashion from effect to cause (the manner of both man and beast) and from cause to effect (the manner of men), thereby anticipating possible future events based on the experiences of the past. When the reasoning faculty is governed by design, the discourse of the mind is but a seeking, similar in fact to the faculty of invention which the ancients call *Sagacitas* and *Solertia*. This is also called remembrance, or a calling to mind (*reminiscentia*), a "reconning" of our former actions. Whereas only the present has a being in nature, and things past have a being in the memory, things in the future have no place at all except for this process of "re-conning" our former actions to anticipate future events. Still, Hobbes cautions that the future is merely a fiction of the mind and cannot be foretold with any certainty; only by applying the sequence of actions past to the actions of the present, done by a prudent man of experience, can the future be foreseen:

> And though it be called Prudence, when the Event answereth our Expectation; yet in its own nature, it is but Presumption. For the foresight of things to come, which is Providence, belongs onely to him by whose will they are to come. (*Leviathan*, 21)

The most certain and sure path to knowledge and truth is to analyze experiences through the memory of present things, which are the only things that have a being in nature. The person with the most experience has the most signs stored in the memory with which to "guess" the future, with the most sagacity and prudence. Indeed, prudence is the presumption of the future, "contracted from the experience of time past." There can only be images of actual experiences in the world of objects; any image, compounded or single, that cannot be referred to the outside world, common to all men, must be eliminated as fancy, the decaying of a dead image in the memory. Furthermore, this sagacity, this wisdom, is acquired not by the reading of books, but by the reading of men and human nature. Therefore, the resemblance in circumstance between Christ sold unto death and the king similarly betrayed must be eliminated from any reasoning process, for the former has no reference beyond the story told in a book, which can neither be verified nor denied.

In this sense, Hobbes takes a more extreme stance on the issue of words and things than Bacon. Whereas Bacon could discover some edifying instruction within the fables of the ancients, indeed finding the reasoning by allegorical displacement and substitution preferable to the philosophizing of Aristotle and Plato, Hobbes severely limits the use one can make of ancient fables, myths, and stories. Even the life of Christ is open to question. In fact, in the final paragraph of chapter 3, Hobbes calls into question the very reality of a personal God. Everything imaginable, he claims, is finite; we can have no mental image of the infinite, there being no object of infinity readily available to the five senses. "When we say any thing is infinite, we signifie onely, that we are not able to conceive the ends, and bounds of the thing named; having no Conception of the thing, but of our own inability" (*Leviathan*, 23). The name of God should not, then, be used to make him more conceivable; rather, that name signifies his incomprehensible nature, his unavailability, his inaccessibility. In order to conceive of anything, that thing must have a place, a dimension, a magnitude, and it must be divided into parts. No other thing is a true conception, merely the absurd speeches of deceived, and deceiving, Schoolmen.

A name is simply a verbal impression of a thing, the image of the object in the mind. For Hobbes, there is no reality (or real knowledge) beyond the words and the mental images we use to represent things. If one wishes to understand, or analyze, a tree, one should investigate that tree as it presses its material substance upon the understanding, then consider how the mind filters, registers, and codes the information. Augustine, in *De Magistro*, would agree with this statement. Yet Augustine enlarges on this idea with a broader conception of the memory. A name, or a teacher using names, triggers something in the pupil's memory, something already there. Hobbes, in contrast, claims that the object itself is matter in motion, and this matter presses, rubs, or strikes the eye, and "makes us fancy a light." These qualities are distinct from the object itself. "For if those Colours, and Sounds, were in the Bodies, or Objects that cause them, they could not bee severed from them, as by glasses, and in Ecchoes by reflection, wee see they are" (*Leviathan*, 12). We see an object—the sun, for example. The sensible quality of this object, its matter, strikes the eye, and we "fancy" it is light. We think it is light, therefore we call it light. Yet, still the object is one thing, and the "image" or "fancy" or "name" is another.[11] All science, then, which Sprat claims is an attempt to make a "faithful *Record* of all the Works of *Nature* or *Art*" (Sprat, 162) is, for Hobbes, but a making of fiction. Furthermore, Hobbes claims that in the universities, scholars follow Aristotle by arguing "that the thing seen, sendeth forth on every side a visible species . . . the receiving whereof into the Eye, is Seeing" (*Leviathan*, 12). This implies that there is some sort of real connection between the object and the perceiver, the thing and the image, fancy or name. Hobbes denies that any such real connection occurs, implying that witnesses, in making their fiction, inevitably fashion a model that positions them within a greater political code (or symbolic order, to use Lacan's phraseology). It is no wonder, then, that Hobbes' epistemology continually calls attention to the political and the partisan, hence, the necessity of a central monarchy to guarantee stability if not truth, and is therefore in strict opposition to the methods and assumptions of Sprat and Boyle.

From this sequence of arguments, Hobbes turns his attention to the issues of speech. The invention of letters, he writes, is the most remarkable feat of mankind, greater even than the invention of printing. This discovery not only allows for the continuation of knowledge, but the transmission of such knowledge from one group of people to another. The most noble and illustrious discovery was the giving of names and appellations, and the subsequent connection made between these names and thoughts. Adam was the first giver of names, instructed by the very speech of God. Unfortunately, this first language has been lost at the Tower of Babel, when, for their foolishness, mankind were deprived of this unifying mode of communication. Henceforth, each tribe would develop their own tongue, distinct from that of the neighboring tribes. This diversity of tongues added to the copiousness present in all languages of the present time, commerce between different groups making assimilation of various words and names necessary.

The use of names is to serve as marks, or notes of remembrance, and to signify by connection and order what a person conceives, desires, and fears. There are four general uses of speech and four corresponding abuses. First, we use speech to register our thoughts so that, by cogitation, we might better understand the root causes of things; this is the acquiring of arts. Second, we counsel and teach others that which we have learned by such cogitation. Third, we make known to others our will and purpose. Fourth, we please and delight ourselves and others by the eloquence and playfulness of our language. The last use, claims Hobbes perhaps facetiously, should of course be innocent, and not used for any serious business or political affairs. This, of course, contrasts with Sidney's defense of the poet as an active, powerful, and necessary actor in the body politic.

The abuses of speech are four in number as well: (1) when we incorrectly register our thoughts by an inconstant signification; (2) when we use words metaphorically, in any other sense than what they literally mean, what they are ordained for; (3) when we declare a thing to be our will, when it is not; and (4) when we use words to cause grief and despair to another, where there is no intention of correction and amendment. Hobbes is concerned that words are used in their most primitive (original) and simple capacity. Metaphor should be shunned where it is not used for some purpose of edification. Obviously excluded here is Hobbes' own metaphorical conceit of the leviathan as an artificial animal created through the mimetic impulse in man, that artificer who copies nature to pattern a well-disposed and well-tempered commonwealth.[12]

Hobbes adds that the consequences of things imagined in the mind are not always identical to the consequences of names. For example, a man born deaf and mute will be able to understand that in one triangle, three angles are equal to two right angles. However, only someone who has the use of words can conclude, and articulate, that universally such equality of angles is in all triangles. Reasoning, then, is a necessary consequence of speech. A particular is registered and remembered, and through experience, becomes a universal law. Indeed, truth consists in the right ordering of names in any affirmation, since true and false are attributes of speech, and not of things. Because of this, the prudent seeker after truth must always remember precisely what every name he uses stands for, and to place it in his scheme accordingly, "or else he will find himselfe entangled in words, as a bird in lime-twiggs, the more he struggles, the more belimed" (*Leviathan*, 28). To do this, he must examine the definitions found in earlier authors in order to correct them, if necessary, or to remember them if the definition is particularly apt. Notice that Hobbes shows little concern for the vivid picture that the word or speech triggers in the imagination; rather, the definition must be clear and simple, and it must always refer to some object the senses can identify in the external world.

It should be noted that, in this section, Hobbes himself builds several analogies to illustrate his point. To become entangled in words is to become entangled like a bird in lime twigs. To become confused trusting the sum of definitions written in ancient authors is to be like the birds, "that entring by the chimney, and finding themselves inclosed in a chamber, flutter at the false light of a glasse window, for want of wit to consider which way they came in" (*Leviathan*, 28). Hobbes' stated purpose is to show the *correct, prudent,* and *reasonable* way of using metaphor, namely to illustrate a point already made in literal prose; the enthymeme, the vivid picture, is drawn directly from the observable occurrences of nature, from the behavior observed in birds. The reader must suppose that humans imitate the birds when they too are trapped by their own ignorance and inability to strengthen their reason. The path to true knowledge is to analyze these metaphors, used by Hobbes and, more important, by the ancient authors, and test their effectiveness in reference to observable reality. Hence the crucial importance of definition:

> In the right definition of names lies the first use of speech; which is the acquisition of science: and in wrong, or no definitions, lies the first abuse; from which proceed all false and senseless tenets; which make those men that take their instruction from the authority of books, and not from their own meditation, to be as much below the condition of ignorant men, as men endued with true science are above it. (*Leviathan*, 28)

Natural, that is, common sense and the imagination are not a matter of absurdity, for nature cannot err. Copiousness, therefore, can make a person either excellently wise or extraordinarily insane. If copious language is pruned and controlled through the proper discovery of accurate definition, then the wise man can seek after truth, knowing that, with his letters, ratiocination will be aided rather than hindered. With improper definitions, however, and a fanciful crafting of metaphor which does not rely on reference or resemblance to nature, confusion, distemper, and madness will occur. While Hobbes had earlier recommended that the reader see these attempts at definition as a matter of fiction-making, later he is concerned that certain fictions are more reasonable and valid than others.

Like the members of the Royal Society, Hobbes' aim is to build a hierarchy of ideas, whereby the ratiocinative ideology that constitutes identity as resemblance to external nature, rather than to fictional myths, is held prior to any system based on fantastical examples found in ancient books. The chief difference between Hobbes and Sprat is that the latter uses external nature as a guarantee of truth regarding the accuracy of scientific and artistic discourse, while the former notes that this guarantee is but a necessary fiction designed for stable control of a civil society. Hobbes' text asks us to perceive rational ideology critically but ends by affirming the inescapable social power and importance of that ideology in coalescing the various fictions through which human life is constituted. For Sprat, the ideological nature of experimental methodology is assumed but remains transparent and invisible through the articulation of a *greater enterprise* to advance *real* (and not fictional) knowledge.

"Words are but wise men's counters, they do but reckon by them." Hobbes in this statement both diminishes and privileges the representational function of language in the construction of knowledge. The fool debases the value of his money, his words, when he derives the definitions from an Aristotle, a Cicero, or an Aquinas. Here, Hobbes makes a play on *accounting* when he notes that ratiocination is called by the "Latines" accounting. In Latin, accounts of money are called *rationes,* and accounting *ratiocinatio.* What the English call *items* (that is, bills or books of account), the Latins call *nomina,* or names. Hence, the Latins extend the word *ratio* to the faculty of reckoning in all other things. The Greeks, in contrast, have only one word for both speech and reason—*logos*—for there is no reasoning without speech.

With Hobbes, the vital bond of speech and reason makes it necessary that words, like things, are properly divided, classified, and reduced to four general heads. First, a thing enters into account for matter or body. Second, a thing enters as an abstraction, being severed not from matter but from the account of matter. Third, a thing enters account as properties of our own bodies. In this instance, when something is seen, we take into account not the thing itself but the sight thereof, the color, the idea as it is registered in the fancy. And, finally, a thing enters which is itself a name or a speech. The reader, and the speaker, must be certain that the words of the speech are not absurd or false. Even our affections, appetites, aversions, and passions are merely conceptions. In the case of these latter passions, the reader (or receiver) of speech, and the speaker must take heed that the diversity of passions and prejudices will be noted so that any speech not be misunderstood or misinterpreted. For this reason, we must take heed of words, "which beside the signification of what we imagine of their nature, have a signification also of the nature, disposition, and interest of the speaker" (*Leviathan,* 32).

For one person, wisdom is called fear; cruelty for one is justice to another. Because of this, names that denote the passions and affections can never be true grounds of any ratiocination. Metaphors and tropes of speech are also excluded from the ratiocinative process. These figures of language, however, are less dangerous, for they announce their inconstancy, their inexact reference to observable reality, whereas the names of passions do not. While the affections are but conceptions in the mind, the names of these affections should not be used as counters in the process of ratiocination. The meaning of the names is so diverse that an adequate reckoning cannot be made. Presumably, the affections have little use in the search for philosophical truth. Hobbes is trying to classify different speech acts. He is not, strictly speaking, abusing the use of metaphor or the use of names signifying moral conduct. Instead, he is appropriating a certain kind of discourse for the act, or art, of ratiocination—namely, a discourse that uses only correct and consistent words.

Hobbes claims that civil war, trouble, and sedition can be resolved by devising absolute linguistic clarity. Feigners, myth-makers, and inventors of fiction have no place in the ideal commonwealth; that is, until they subtract from their words, their names, any fanciful and dreamlike references. In Hobbes' political theory the commonwealth must be understood and instituted using language that has been freed of fanciful, "affective" and emotional associations. For Hobbes, reasoning is the second stage in the process of ratiocination. The first step, as we have seen, involves the imposition of definitions on words and names so that the object named can be placed most effectively in the memory. Reasoning involves discarding improper names and providing connections between proper definitions so that a general assumption can be maintained. This event is called science; unlike sense and memory, it is not inborn but attained by industry and labor. Reason is a reckoning, an adding and subtracting, of the consequences due to the imposition of names in ratiocination. While names and words are the counters whereby things are reckoned and taken into account, the process itself is called reason. Here, Hobbes distinguishes between marking, when the process is solitary, for the purpose of one person, and signifying, when the process is demonstrated to other men. The end, and use, of reason is not to find the sum of all knowledge or to discover a complete truth. Rather, one must start with settled signification or definition of names and proceed from one consequence to another, there being no certainty of the last conclusion without a certainty of all the intervening affirmations.

In the reasoning process, the seeker or laborer must take heed not to fall into the absurdity of philosophy. Among these absurdities is the want of method, definitions, and settled significations of words. Another absurdity is the imposition of a name that signifies nothing. The schools are notorious for imposing such words as hypostatical, transubstantiate, consubstantiate. The final absurdity is the use of metaphor, tropes, and other rhetorical figures instead of words proper. In "reckoning, and seeking of truth, such speeches are not to be admitted" (*Leviathan*, 36).

Children are not endowed with reason at all until they have mastered the art and use of speech. Similarly, some people are not equipped to seek truth, to reckon properly, to engage in science and geometry. However, their natural prudence will ensure that they will conduct themselves within the state and govern their actions with temperance and virtue. In fact, these individuals are in a better situation than their more nobly conditioned "brothers" who, by misreading and trusting "wrong" reason, fall into error, believing false and absurd rules rather than laboring to achieve true and settled significations. Trusting in the authority of books is to follow the blind man blindly.

> To conclude, the light of human minds is perspicuous words, but by exact definitions first snuffed, and purged from ambiguity; reason is the pace; increase of science, the way; and the benefit of mankind, the end. And, on the contrary, metaphors, and senseless and ambiguous words, are like *ignes fatui;* and reasoning upon them is wandering amongst innumerable absurdities; and their end, contention and sedition, or contempt. (*Leviathan*, 37–38)

Hobbes, however, provides the provision that this knowledge of consequence, which is also called science, is not absolute but conditional. No discourse can end in absolute knowledge of the past, the present or the time to come. In fact, signs of prudence are all uncertain, since to observe, experience, and remember all things is impossible. This notwithstanding, though natural judgment be not infallible, one should not then seek truth in authors. Those who do so will not advance in politics or history, but merely shine and polish the reputation of their own wit and conceit.

In his work on grammatical theory in the seventeenth century, G. A. Padley argues that the distrust of words and fascination with nomenclature displayed generally by thinkers and writers throughout the century can be traced back to medieval nominalism.

> This . . . trait links up again with the utilitarianism of contemporary puritans, many of them Ramists, with their emphasis on a plain preaching style and their preference for the world of "things." Particularly among the new men of science, who in this join hands with the puritans and with pedagogical reformers such as Comenius, there is a desire to achieve contact with reality and use it to practical ends. It is this that lies behind the contemporary longing for a system of communication that would be free of the treacherous ambiguities of natural languages, and in which concepts (by definition the same in all men) would each have their single,

unalterable sign. In order to achieve this, the universe had first to be cut up into its constituent parts, analysed and labelled, and it is here that the continuing Scholastic influence becomes paramount. (Padley, *Grammatical Theory in Western Europe* 1:325–26)

We have seen how Hobbes tries to divide the world into its many constituent parts, then find the appropriate label with which to define these parts. This kind of mentality can also be seen in the several schemes in the seventeenth century regarding the devising of a universal language, a "real character" or alphabet which would give the exact and appropriate iconic notification of "things" in every category of experience.[13]

In England, these programs for an artificial and universal language are based on an empirical methodology derived from Bacon which has at its center the notion of "sense-realism" (Padley, *Grammatical Theory* 1:327). Bacon's "instauratio magna" proved extremely influential to the new science of the seventeenth century, as figured in the Royal Society. For example, John Wilkins, in his *An Essay Towards a Real Character and a Philosophical Language,* credits Bacon as the primary inspiration of his work. In the spirit of Hobbes and Sprat, Wilkins claims that by reading this dictionary the great wits of all nations will not only polish their own language, but also become aware how things are to be preferred to words, just as real knowledge is beyond elegancy of speech and the general good beyond the particular good of any nation.[14] In addition, "this design will likewise contribute much to the clearing of some of our Modern differences in Religion, by unmasking many wild errors, that shelter themselves under the disguise of affected phrases" (Wilkins, "Dedicatory Epistle," i). Instead, Wilkins aims to determine the genuine and natural importance of words, thereby revealing the opaque and flat nature beneath several enigmatic notions that are otherwise articulated in a great swelling style.

In devising a real character taken from the natural notion of things, Wilkins draws up "the Tables of Substance, or the species of Natural Bodies, reduced under their several Heads." He continues with other tables of accidents, which are further reduced and changed so that all simple things and notions are considered a priori. In this system of accounting, all extraneous ideas that are attached to words are removed; the common and general nature of things is best expressed by a specific, narrowly defined and classified word. In his approach, Wilkins reinterprets Bacon's statement that real characters should "coincide in number with the radical words of his ideal language . . . and to this end he opposed to Aristotle's theory of art as the imitation of nature his own theory of the 'congruity of natural and artificial phenomena.'"[15] Language, then, becomes the mirror of the universe, mediated by the understanding which notes, marks, and then reflects on the material world.

Both Hobbes and Wilkins express confidence in language as an accurate mirror of the universe, reflecting as it does the congruence between the intellectual world of the understanding and the concrete world of matter. Bacon, of course, does not have such complete faith in the ability of any language, even that of hieroglyphs and "real characters," to mirror precisely the things of nature. After all, the idols of the marketplace, the misreadings of language by the populace, ensure that the mind is a "false mirror which distorts the nature of things by mingling its own nature with it" (*Novum Organum*, aphorism xli). As such, it is often misleading to find the inspiration of a universal language or a real alphabet in the "instauratio magna" of Bacon.[16]

Despite this qualification, one can easily see the influence of Baconian ideas in the work of mid-seventeenth-century theorists on language who were struggling to resolve the problem of verbal ambiguity. Robert Boyle, member of the Royal Society and follower of Baconian empiricism, recognizes the difficulties of expressing spiritual thoughts using scientific, empirical methods. Boyle writes that apposite comparison, similes, continue "not only (to) give Light, but Strength, to the passages they belong to, since they are not always bare Pictures and Resemblances, but a kind of Argument; being oftentimes, if I may so call them, Analogous Instances, which do declare the Nature, or way of operating, of the Thing they relate to, and by that means as in a sort prove, that, as 'tis possible, so it is not improbable, that the thing may be such as 'tis represented."[17] As such, he feels free to draw comparisons from telescopes, microscopes, and so on, for the purpose of teaching and delighting his audience, to make the notions they convey better kept in the memory:

> Proper Comparisons do the Imagination almost as much Service, as Microscopes do the Eye; for, as this Instrument gives us a distinct view of divers minute Things, which our naked Eyes cannot well discern; because these Glasses represent them far more large, than by the barre Eye we

judge them; so a skilfully chosen, and well-applied, Comparison much helps the Imagination, by illustrating Things scarce discernible, so as to represent them by Things much more familiar and easy to be apprehended. (*Christian Virtuoso,* A6)

The experimental scientist is compared to a diver who must fetch the pearls, corals, and other precious things that lie buried on the bottom of the lake, not content to fetch those things that lie on the surface of the sea. In this, Boyle follows the lead of Bacon, the illustrious Verulam, who, though not a florid writer, frequently makes use of apt comparisons.

An apposite comparison—that is, a simile or metaphor—is not simply an alternative or ingenious way of expressing a particular thought which might, just as effectively, be uttered in a plain or literal way. Rather, many principles, whether they be rational thoughts or emotional states, require an analogical imagination, and therefore an analogical style of discourse, to be grasped and expressed. Boyle, unlike Glanville, is disturbed by this loss of analogical thought in language. As such, one might well place him outside that group of Royal Society scientists which felt a certain optimism in the ability of plain words to express or, rather, to "mark" obvious and self-evident truths. This clear correspondence between word and thing, while, for many Restoration writers, offering a satisfying solution to the problem of how language signifies, proved less than adequate for such writers as Boyle. For him, words possessed an importance that went beyond the notions of transparency in sign and signified articulated by Hobbes and Wilkins.

Throughout his career as a writer and scientist, Boyle was concerned about the motivation of the scientist and the philosopher. He was constantly trying to create an audience for natural philosophy; as such, his writing sought to persuade, as well as to construct truth, an epistemological stance that linked him with Bacon. Yet, there was still the sense that, for Boyle, language could change the subject matter. In *Certain Physiological Essays,* Boyle claimed to write in a somewhat philosophical (as opposed to a rhetorical) strain, so that his expressions were clear and significant and not merely curiously adorned.[18] The style, however, must not disgust the readers by its flatness,

> especially when he does not so much deliver experiments or explicate them, as make reflections or discourses on them.... For on such occa-

sions he may be allowed the liberty of recreating his reader and himself, and manifesting that he declined the ornaments of language, not out of necessity but discretion, which forbids them to be used, where they may darken as well as adorn the subject they are applied to. (*Works,* i, 304)

This idea sounds very much like the ideology of Erasmus, who also urged the writer to be re-created, to become the text. Because of this ability of writing to change the reader, the scientist should not darken the subject, but should, in fact, change, alter, or otherwise construct the subject so that the reader can be properly educated.

Boyle constantly showed an interest in how the material would be presented or laid out on the page. In a letter to Boyle, John Beale notes that "every marginal reference might advertise of a fresh method by union, parallel, gradation, limitation, and thousands of other inferences, whenof some might prove of comprehensive and pregnant importances" (*Works,* vi, 404–5). Experiments would be accompanied by illustrations, by letters, dialogues, reports, descriptions, sometimes even quotations from sermons, meditations, and oratory. As such, Boyle seemed attuned to the artificiality of his discourse. All these devices proved to be a useful extra intended to manipulate the reader's response and mesmerize, almost seduce, the audience with the validity of the particular truth claim.[19] In many of his dialogues, Boyle adopted a persona he called Carneades, a Boyle yet not a Boyle. Again, like More and Erasmus, Boyle wished to make truth claims valid, not a matter of political propaganda.[20] For Boyle, how the text behaved became the main subject matter of the text. Language was not transparent or referential; rather, for Boyle, language was to be an instrument for the scientist's unmediated cognitive and experimental grasp of reality. Metaphors were to the mind what microscopes were to the eye, re-creating *in a changed form* the natural phenomena of the world.

If language was like a telescope, however, how then could the scientist or group of scientists determine which instrument was faulty, which was true or valid? Hobbes, like Descartes, validated the efforts of the individual natural philosopher. In contrast, Boyle emphasized the collaborative efforts of numerous investigators, the corroborated evidence and matters-of-fact of eyewitness testimony. Like Bacon, Boyle relied on the practices and procedures of common-law jurists. Eyewitness accounts would be used to verify results; in this scheme, the comparison of two corroborating accounts would validate the reality represented, as well as the human minds that constructed the testimony.

Hobbes, in contrast, championed the analytical approach typical of Roman law, opposing the experimental science and favoring a rational and mathematical approach instead. Boyle, in contrast, relied on common law, with its more historical, empirical approach. The common law, which required many long years of study and experience, seemed to reflect Boyle's attitude that the "experimenter's way" was neither short nor easy, that experiments were troublesome to make. The road to becoming a reliable witness was long and arduous; signs of reliability would include experience, discretion, a balanced (indifferent) personality, a lack of "enthusiasm" or passion. The experimentalists would be like the common-law attorney, using artifical reason (not dialectical analysis), asking how precedent should apply to present cases, and relying upon "moral demonstration." Trials (experiments) would be a public affair; witnesses would testify in an open court. A jury of twelve men would deliver their verdict on the facts of the case. Complete proof would not be needed; the case would be decided in favor of the person whose account appeared to be the most likely to be true. As Rose-Mary Sargent notes, "the 'proof' of the case ... consist(s) in the 'finding of a body of reasonable men' according to the probabilities of the case."[21] This epistemology relied on the expert witness weighing the evidence.

There is evidence in Boyle's work that he did indeed use the methodology of the common-law jurist when reflecting on the issue of scientific verification. In his work, *Things Said to Transcend Reason,* Boyle asserted that the nature of those things not capable of mathematical or metaphysical demonstrations, and yet being really truths, "have a just title to our assents ... it being sufficient that they are strong enough to deserve a wise man's acquiescence in them" (*Things Said to Transcend Reason, Works* 4:449). Boyle also relied on the common law's idea of probable cause, that is, the absence of specific reasons to doubt. "Though each testimony single be but probable, yet a concurrence of such probabilities may well account to a moral certainty" (*Some Considerations about Reason and Religion, Works* 4:1).

My final question concerning Boyle has to do with the nature of these perfect instruments, or witnesses, that discover hidden processes of nature. In a 1647 letter to Samuel Hartlib, Boyle states his admiration for the Hollanders and the Venetians, whose ingenuity was courted with the greatest encouragement. Boyle praised the continental commonwealths that promoted talent, a work ethos, and public service.

> Certainly the taking notice of, and countenancing men of rare industry and publick spirit, is a piece of policy as vastly advantageous to all states, as it is ruinously neglected by the most. And therefore we may evidently observe those commonwealths (as the Hollanders and the Venetian) to be the most happy and the most flourishing, where ingenuity is courted with the greatest encouragements." (letter to Samuel Hartlib, May 1647, in *Works* 1:xl)

Curiously, this attitude is linked to a Protestant and capitalistic work ethic, as well as an Erasmian code of civic virtue that Boyle surely inherited from his father, the first earl of Cork who was himself a capitalist speculator in Ireland.

According to Nicholas Canny, Richard Boyle passed along several important ideas to his son, which explains Boyle's preoccupation with the connection between religion and politics.[22] These ideas include "the belief that in pursuing practical goals he was fulfilling God's purpose; his belief that the pursuit of evangelical work among a primitive people was a futile exercise until such time as they had been incorporated into a disciplining social framework; his belief in a broadly based Protestant church 'that would tread a middle course between the extremes of sectarianism and strict conformity'" (Canny, 147). Charles Webster's thesis is that Boyle's family, especially his father and sister, the Viscountess Ranelagh, were in frequent contact with a small group of Protestant refugees, primarily from the Platinate, Bohemia, and the Netherlands, who espoused millenarian views that linked a radical Protestant view of God's providence and plan with the needs of practical scientific inquiry.[23] The London home of the Viscountess Ranelagh became an important meeting place for a group of Irish refugees and exiles, including puritan divines and scientists. In this group were Robert Boyle, his brother Roger, the Baron Broghill, Sir John Clotworthy, and Archbishop James Ussher. Initially created as an autonomous group independent of the Hartlib Circle, the group was known as the Invisible College (Canny, 142).

J. R. Jacob contends that Cromwellian Ireland became the resort of scientists of the Hartlib stamp, though independent of that group, because that country was reduced to ruins first by the Irish rebels, and then by the Cromwellian army.[24] This political vacuum allowed for the creation of a completely new society based on scientific principles. According to Canny, the "nub of (Jacob's) thesis is that English puritans, and most especially the members of the Hartlib Circle, indulged in a selective reading of the writings of Francis Bacon which enabled them to dovetail his views on scientific enquiry into their own millenarian prognostication" (Canny, 200).[25] In fact, the primary tenet of Hartlibean science was its stress on "gaining control of the physical environment through improvement in technology" (Canny, 148).

These Irish Protestants were always afraid that their reforms would be swept away by Counter-Reformation ideas; therefore they resorted to millenarian expectations that convinced them they were the chosen people. English literary writers who encouraged this view included Edmund Spenser, Sir John Davies, and William Herbert. Like these writers, Boyle's father, and perhaps Boyle himself, was inspired by the providential model of Florentine virtu, whereby the ingenious scientist turned the inexplicable and unpredictable nature of Fortune to his advantage. As such, the notion of providential intervention in human affairs actually served as a stimulus to scientific investigation, and it goes a long way toward explaining Robert Boyle's insistence on a certain type of reliable eyewitness investigator, a moderate gentleman Christian virtuoso whose word could be corroborated and believed by his peers. In conclusion, Boyle's aim, like Spenser's, would be to "fashion a scientific gentleman" as a reasonably accurate instrument to weigh knowledge, and to persuade the reading public that this expert witness (instrument) would be reliable, capable of providing a moral demonstration similar to that used in the common-law courts of England—that is, that there is an absence of specific reasons to doubt.

# 6
## The Figure in the Pool: Milton's Epistemology of Nature

> I used to tell my friends that the inventor of painting, according to the poets, was Narcissus, who was turned into a flower; for, as painting is the flower of all the arts, so the tale of Narcissus fits our purpose perfectly. What is painting but the act of embracing by means of art the surface of the pool?
> —Leone Battista Alberti, *De Pictura,* Book II

IN CHAPTER 5, I PRESENTED VARIOUS SCHEMES OFFERED BY members of the Royal Society to resolve the conflict of how one can properly translate things into words. M. M. Slaughter agrees that "in the seventeenth century it was generally felt that natural language was not isomorphic with nature or reality—that words and things did not properly match. . . . All language needed to be rearranged, readjusted, and developed . . . it was imperative that the scientists create nomenclatures—compilations of words or terms which nominated or designated things."[1] Such scientific taxonomy would become a method which established the orders of nature, and substantiated such notions that nature was constructed in a regular way, that nature could be known and described, that objects of nature would be classified in a systematic way, such as in *genera* and *species,* that words might be isomorphic to things, and that definitions based on such divisions ultimately became the essence of things in scientific belief. Scientists (natural historians) and linguists both attempted to invent a "linguistic system in which things were simultaneously designated and defined."[2] Vision, the ability to apprehend and comprehend the essence of natural phenomena, became critically important. Along with this preoccupation with naming comes an obsession with accurate visual perception. Galileo's experiments with the telescope, as well as Descartes' and Pascal's work on optics, testifies "to the seriousness of the scientific interest in visual phenomena, and particularly to the determination to find 'truth' in the visual apprehension of the world."[3]

In his *The Curious Perspective,* Ernest B. Gilman writes that "linear perspective, which begins in the earlier Renaissance as an expression of confidence in the certainty of human knowledge, gradually comes to reflect a more complex and ambiguous relationship between the knower and the knowable."[4] While Giotto, Raphael, and Alberti were concerned with ideal space, obeying the classical rules of symmetry and proportion, in the seventeenth century the more sophisticated use of lenses, the telescope, and the microscope opened up a new space, and a new world, to scientists and artists. The stable, secure point of view of the Renaissance had given way to the more ambiguous, problematic view of the Baroque. Both the viewer and the reader must figure out their own relationship and engagement to the work of art in question. Ironically, the period's preoccupation with mathematical certainty and truth nurtured a type of art (anamorphic art based on distortion) that was epistemologically opposed to the kind of truthful representation of visible forms inherent in traditional definitions of *mimesis.*[5]

> By locating the unique viewpoint in nontraditional way—for example, far to one side and not on a line crossing the painted surface—the anamorphic artist reminds the viewer of the active role that the viewer must take in deciphering the image. . . . In this way anamorphoses . . . force on the viewer an attitude of obedience, although anamorphic art, unlike baroque art generally, instills in the viewer the consciousness of the viewer's individual activity.[6]

Lyons argues that anamorphic art invites the viewer to contemplate the work from only one viewpoint: creating images that are incomprehensible unless seen from that narrowly defined perspective. Lyons also notes that anamorphic art reminds the viewer that an active role must be assumed to decipher the picture, a role that is ultimately denied when the correct posture, or reading position, is determined. The artist's task is to teach the reader how to find this proper position. Distortion and illusion are deliberately highlighted in an effort to teach the reader how to move beyond and away from these visual distractions to find the real and the true, reminding the viewer "of the distinction between appearance and reality."

This seventeenth-century preoccupation with anamorphosis and the related concern with visual perception addresses the multiple problems regarding the sign (the picture), the signified (the reality), and the connection between the two. For the Jansenists, conceiving of thought as a mental picture has dangerous consequences.[7] Since we are able to conceive of "a very great number of things of which we have no image at all," the imaginative faculty must be made subordinate to the pure intellect.[8] As such, it is false to suppose that all our ideas come through the senses, or have a real connection with any bodily image. It follows that the Jansenists deny the importance of metaphor in producing vivid images in the intellect. We can express nothing by our words without having a clear, self-evident idea already present in the pure intellect without words or images derived from sense perception.

Pascal calls this kind of reasoning *l'art de géométrique* and distinguishes it from the arts of persuasion. In the former, only nominal definitions are admissible, "the arbitrary attribution of a name to things which have been clearly designated in perfectly known terms . . . so that the chosen name remains empty of all other meaning."[9] The goal in geometrical reasoning is not to define objects or prove principles, for this kind of reasoning has no need of such processes: "the cause which makes them incapable of being demonstrated is not their obscurity, but on the contrary their extreme clarity . . . which convinces reason more forcefully than does argument." Indeed, nature is entirely familiar and common, and does not need the mind to be "hoisted" to some elevated understanding. "Strained and labored manners fill it with a foolish presumption by an unnatural elevation and a vain and ridiculous inflation."

> One of the chief reasons why those who enter into this knowledge of the right road that ought to be followed are diverted is the opinion that they set forth at the outset that the good things are inaccessible, by giving them names like "great," "lofty," "elevated," "sublime." Everything is lost this way. I would like to call them "low," "common," "familiar."[10]

My aim in this chapter is to examine the poetry of Milton in light of these ideas.[11] The central question involves how this poet used language to draw the mind away from the confusion and obscurity inherent in any figurative and metaphorical utterance, finding a transparent style that would reveal the naked truth of things.

In Book V of *Paradise Lost,* Raphael tells Adam that "what surmounts the reach / Of human sense, I shall delineate so / By likening spiritual to corporal forms" (V, 571–73). Yet, this kind of discourse merely addresses "discursive reason," the *ratio scientiae,* which is related to objective reason and is used by men to build a theory of the laws of nature based on the facts of primary perception. The angels, who exist on a higher intellectual and spiritual plane, and are thereby closer to the reasoning faculty of God, do not need such a language as a means of communication. Unfallen as they are, the angels rely on intuition, a "simple and undifferentiated operation of the contemplating intellect." Through this intuitive faculty, the angels are able to pursue knowledge without language as a discursive means.

This subordination of discursive argumentation and dispute is addressed by Thomas Sloan in his study of Donne, Milton, and the decline of humanist rhetoric in the seventeenth century.[12] Sloan argues that both Donne and Milton, though exposed to and trained in Ciceronian rhetoric, lived and wrote in a period that saw the disintegration of such rhetoric. Donne, trained as a lawyer, relied more heavily on the methods of *controversia.* In contrast, Milton's epistemological stance rested on an alternative mode of composition that was non-lawyerly and noncontroversial, namely Ramist. In his *Art of Logic,* Milton shows his preference for the methodology espoused by the Ramists, a logic that, though rooted in rhetoric, elevated the art of thinking to a position dominating any rhetorical arts of disputation or argumentation.[13] In the logic of Peter Ramus, there is little room for the kind of controversial thinking that so appealed to the humanist rhetorician. Rather, Ramism tended to simplify logic, or dialectic, by limiting it to *inventio* and *dispositio,* and relegating rhetoric to mere ornamentation. To dispute and argue in terms of pro and contra, to utilize the vital colors of rhetoric, and to access that storehouse of the memory which is the orator's primary task, has no place in Ramist logic. Instead, Ramism uses simple diagrams or thought patterns, removing the voice entirely from the reasoning process. The silent intellect became the seat of reason. Whereas rhetorical argumentation, arising from and proceeding by antithesis, tends to be open-ended, supple and flexible, with the synthesis proving to be less absolute, logical argumentation is considerably less negotiable. Ong has commented that Ramist logic, more than other method of formal logic, is "extraordinarily assertive and autocratic," predicating an unusual confidence in its own single-minded convictions.[14]

Ong acknowledges that Milton's erudition and wide reading ensured that his logic lacked the utter simplicity inherent in the work of lesser Ramists, marking a certain Ciceronian and rhetorical cast to his writing. Nevertheless, "Milton remains, through his *Logic* and its relationship to his other work, linked to Ramism more clearly than any other poet of much stature."[15] Sidney, the other major poet associated with Ramism through Abraham Fraunce's *Arcadian Rhetorike*, is still part of that Erasmian humanism which seeks access to truth through the various protean shapes that language, particularly poetic and oral language, can assume. In contrast, Milton shows a greater wariness in face of such virtuoso verbal feats. Ironically, Milton the poet shows as profound a distrust of language, of words, as Bacon the experimental scientist. In *The Reason of Church Government Urged Against Prelaty*, Milton articulates the stance that is by now a familiar seventeenth-century refrain.

> But my hope is, that the people of England will not suffer themselves to be juggled thus out of their faith and religion by a mist of names cast before their eyes, but will search wisely by the scriptures, and look quite through this fraudulent aspersion by a disgraceful name into the things themselves.[16]

In this sense, Milton follows the linguistic ethic of Thomas Hobbes which relates faulty (or false) thinking with the imposition of vague and unsuitable names.[17] In the same chapter, Milton charges these prelates with sophistical argumentation which seduces the weaker sort of people, and then confuses them by scandalous misnaming, "thereby inciting the blinder sort of people to mislike and deride sound doctrine and good Christianity, under two or three vile and hateful terms."[18] Hobbes, we have seen, is forever warning the reader that the fantastical stories and fables of the poets are best linked to the inferior faculties called the fancy, rather than the higher and more reliable faculties of the understanding. Similarly, Milton contrasts a naked and unblemished right reason with the "false-whited and lawny resemblance of truth" which is offered by the prelacy, "like that airborn Helena in the fables, made by the sorcery of prelates."[19]

> And thus prelaty, both in her fleshly supportments, in her carnal doctrine of ceremony and tradition, in her violent and secular power, going quite counter to the prime end of Christ's coming in the flesh, that is, to reveal

> his truth, his glory, and his might, in a clean contrary manner than prelaty seeks to do, thwarting and defeating the great mystery of God; I do not conclude that prelaty is antichristian, for what need I? the things themselves conclude it. (*Reason of Church Government*, 500)

The gaudy, unnecessary embellishments of the prelates are figured in the words "fleshly supportments," "carnal doctrine," "ceremony," and "tradition." These are set against the clean, contrary manner of Christ who appears before the world in his naked flesh, in his truth, glory, and might. As such, Milton need not dispute, argue, or prove the truth of this claim. Rather, the thing itself concludes the argument as a self-evident truth to the seasoned and virtuous mind.

Stanley Fish sees in Milton's strategy, displayed throughout the text of *Reason of Church Government*, a subtle, consistent dismantling and subversion of reason as a movement of the mind. Ratiocination, as arguments formally set down, is opposed to intuition which "implies itself" as a self-evident truth. Milton's Ramism is thus displayed in its autocratic omission of process, dispute, and dialogue leading to consensus regarding probable truth. There is no room for probability in the thinking, and therefore in the stylistic strategy of Milton. While Milton ostensibly structures his discourse with an eye toward logical sequences, frequently using explanatory and qualifying phrases (for example, "therefore" and "but") and dividing and arranging the text into books and chapters, ultimately he negates this sense of reason, logic, and law in favor of God's command and fiat.

Fish explains that, by doing so, Milton establishes two classes of readers/observers, those whose reason has been seasoned by God and those whose intellect has not, the latter being the "unelect" and the unregenerate. The seasoned, elected reader is thereby forced to abandon discursive structures and confront the object—that is, God and his commands—nakedly and in solitude. "The choice-but-no-choice these sentences proffer in their different ways amounts to saying (again and again), 'For God's sake (and your own) see things as they really are' without the mediating and distorting screen of 'coverings,' 'outsides,' and 'reasons formally set down,' all of which are finally equivalent."[20] Just as the ceremonies of the church are unnecessary hindrances in understanding God's commands, artificial and ornate language is unnecessary in the articulation of truth. In fact, there is a sense that writing generally is unnecessary, and must be effaced within that intuitive moment when the reason is illuminated.

What, then, is the purpose of Milton's prose, his stylistic strategy which, in the name of right reason, subverts a discursive process that should invite argumentation to proceed from cause to effect, from thesis to antithesis to synthesis? Again, I refer to Fish for one possible solution. Milton, he writes, hopes that the reader will peer through the mists of words into the things themselves, nakedly presented to the prepared mind. By "words," Milton means the outward show of things, the embellished trappings and dress of thought. Yet this outward show includes not merely ornament and false semblances, but rational argument proceeding from discursive, logical habits. Rather than represent a clear ratiocinative progress, Milton's strategy aims at "a progressive enlargement of the understanding . . . accumulating and cross-referencing pairs of coordinate words and images on either side of a great divide."[21] The carnal is paired and contrasted with the spiritual, the varnished with the plain, darkness with light, polluted with pure, the hidden with the open. The space between the two contrasting images is not negotiated through any discursive procedure or method, as in Bacon's hunt of Pan, the literate experience, but through sheer intuition of a dogmatic faith. Ultimately, the grandeur of Milton's prose requires the reader to acquiesce in the author's identification of carnality, illusion and veiled idolatry with the prelate, and spirituality, plainness and open truth with the presbyter. Dialectic (Milton's logic) thereby becomes propaganda as it loses its final claim to discursive, logical habits of disputation and argumentation, the *controversia* of humanist rhetoric.

Beginning in the second book of *The Reason of Church Government,* Milton presents himself as a second Homer, adorning his native tongue "not to make verbal curiosities the end (that were a toilsome vanity,) but to be an interpreter and relater of the best and sagest things among mine own citizens throughout this island in the mother dialect" (*Reason of Church Government,* 478). He considers the various options open to him in choosing a manner of writing, pondering the choice between imitating the epics of Homer, Virgil, and Tasso, following the strict rules of Aristotle, mimicking the dramas of Sophocles and Euripides, or the pastoral drama of the Song of Solomon. Milton concludes that time is ill spent considering the choices offered here, for the worth of ancient writings lies not in their divine argument, but in the very critical art of composition, in the poetic ability which is the inspired gift of God and which empowers the artist "to imbreed and cherish in a great people the seeds of virtue and public civility, to allay the perturbations of the mind, and set the affections in right tune; to celebrate in glorious and lofty hymns the throne and equipage of God's almightiness" (*Reason of Church Government,* 479).

The writer must paint over and describe all the instances of example that will teach the whole book of sanctity and virtue. As such, the true poet must be someone who is moral and decent, and not of the libidinous and ignorant poetasters who scarce have ever heard of true poetry. For Milton, then, the true poet who embarks on the true understanding of things, imposing a right and faithful correspondence between words and things, language and reality, must be sanctioned and seasoned by God, much like Moses the lawgiver, who instructs his people and informs their understanding by showing in all clarity the *just demands* of God. In this scenario, the assent of the reasoning mind is superfluous, argumentation and dispute unnecessary. Right reason becomes the intuition of self-evident truths.

While acknowledging that other methods of acquiring and transmitting knowledge are useful and enriching (namely, the rules of Aristotle and the rules of nature), Milton ultimately denies the utility of these avenues. He writes that "time serves not" to know which method is the superior one; indeed, he calls not upon dame memory and her siren daughters, but relies upon devout prayer "that eternal Spirit, who can enrich ... all utterance and knowledge." To this is added "industrious and select reading, steady observation, insight into all seemly and generous arts and affairs." With these tools, Milton will leave the pleasing solitariness of his study, where he can in leisure study the bright countenance of truth, to "embark in a troubled sea of noises and hoarse disputes ... to come into the dim reflection of hollow antiquities sold by the seeming bulk, and there be fain to club quotations with men whose learning and belief lies in marginal stuffings" (*Reason of Church Government,* 481). This certainly echoes the intention of Augustine. However, Augustine acknowledges, more than Milton, the necessity of rhetorical tools in this enterprise.

Bacon made a similar claim, or threat, to leave the private sanctuary of study and embark upon an ambitious reform of learning. He too denied the intellectual disputation of the academicians, the schoolmen following Aristotle and the humanists following Plato. However, Bacon did not retreat into prayer and intuitive reason, negating the fund of wisdom collected in the prerational antique fables. Instead, he denies the possibility of limits and fixed boundaries, finding in myth an apt model for his own "restauration" of knowledge. A science based on a true induction requires a mind and an attitude that can move well within fluid and ambiguous boundaries and meanings. Such a mentality is the outstanding feature of the mythological world.[22] In contrast, for Milton this shiftiness is suspect; myth is incorporated into a larger and more stable design, a moral pattern that accommodates myth within the realm of Christian values and theology.[23]

This stability is achieved by a clear dichotomy of the values discussed above, namely, the opposition between the ornamental and the plain, the carnal and the spiritual, the real and the imitation. These juxtapositions are made clear within the moral register of *Paradise Lost*.[24] Resemblance is almost always used in reference to Satan, and his need to put a false appearance onto his own nature. In the second book of *Paradise Lost,* Satan and his council debate about the nature of their battle with heaven, and the extent to which their own hell resembles the celestial home of the unfallen angels. Belial's words are clothed in reason's garb, as he advises ignoble ease and peaceful sloth in hell rather than continued combat to retrieve a place in heaven. Likewise, Mammon prefers to remain separated from the celestial angels, finding in hell a resemblance to the dark nature of God.

> How oft amidst
> Thick clouds and dark doth Heaven's all-ruling Sire
> Choose to reside, his Glory unobscured,
> And with the majesty of darkness round
> Covers his throne, from whence deep thunders roar
> Mustering their rage, and Heaven resembles Hell?
> As he our darkness, cannot we his light
> Imitate when we please?
> (*Paradise Lost,* II, 264–70)

God, in this instance, is the reality and the light; the most that Mammon and his fellows can hope for is a base imitation. Also notable is Mammon's assumption that God's darkness, his majesty surrounded by thick and darkened clouds, is but an imitation of their own nature. Indeed, this darkness in no way hinders or obscures God's majesty, and thus seems to justify the darkness of the fallen angels. The mirror motif is inherent in the thinking and the linguistic strategy of Satan and his followers. This resemblance is in itself a near identification; in the mind of Mammon the copy is identical to the original and is not merely a superficial appearance. Indeed, Satan uses this attitude towards resemblance to establish himself as their mighty "paramount" and "dread emperor," justifying his reign "with pomp supreme / And god-like imitated State" (II, 510–11). Satan becomes an idol of majesty, surrounded by flaming cherubim and golden shields.

In contrast, resemblance in heaven has no real place or function. Later, in Book V, the word *resemblance* is used when describing the starry sphere of the planets, "fixed, in all her Wheels / Resembles nearest; mazes intricate / Eccentric, intervolved, yet regular / Then most, when most irregular they seem" (V, 621–24). In heaven, semblances are not mistaken for the real thing, as they are in hell; rather, the idea of resemblance is continually set against the notions of "seeming," of a near identification that should not be taken for truth. The eccentric motions of the planets seem irregular, yet there is a regular pattern that underlies their eccentricity. The mind must not be fooled by the illusions of perception where likenesses justify a false interpretation of reality. The only function of resemblances in heaven is to entertain; in mystical song and dance the spheres resemble intricate and eccentric mazes that innocently perplex and amuse the mind. Resemblances and analogies can also be used to teach, as when Raphael narrates the war in heaven to instruct Adam on the justice of God, and the secrets of another world. Since these truths surmount the reach of human sense, Raphael must, for Adam's good, resort to "likening spiritual to corporal forms / As may express them best" (V, 571–74). Earth is but a shadow of heaven, and the reality of one is impressed, as a near identical image, onto the other. Yet, the two, heaven and earth, are like each other, not identical, Raphael reminds Adam.

Outside the unfallen realm of heaven, imitations and resemblances take on a more sinister cast. In Book VI (lines 114–26), Abdiel remarks on the unusual power of resemblance that remains when faith and the sense of "reality" have disappeared. To the eyes, to sight, Satan is unconquerable, unchanged in strength and might. Even without the Almighty's aid, Satan appears a formidable opponent, who might yet win in force of arms, just as he in the past has won debates of truth despite his false and unsound reason. This resemblance, though divorced from reality and faith, is still a force to be reckoned with. Indeed, in Book IX, Satan begins his temptation of Eve by calling her the "fairest resemblance of thy Maker fair" and proceeding to identify her with God. All things, by gazing upon her with ravishment, adore and admire her. "A Goddess among Gods, adored and serv'd / By Angels numberless, thy daily Train" (IX, 547–48). By so identifying Eve with God, where resemblance becomes the original, the real, Satan makes an idol of her, and a counterfeit. Eve is likewise encouraged to look at herself in awe and admiration, finding in her mirrored reflection the image of the Creator. This happens early for Eve. In Book IV, she tells Adam that while gazing with love and desire at her own reflection in the lake, she came by her first "unexperienced" thought.

> As I bent down to look, just opposite,
> A shape within the watery gleam appeared
> Bending to look on me, I started back,
> It started back, but pleased I soon return'd,
> Pleased it returned as soon with answering looks
> Of sympathy and love; there I had fixed
> Mine eyes till now, and pined with vain desire,
> Had not a voice thus warned me: "What thou seest,
> What there thou seest fair Creature is thyself;
> With thee it came and goes: but follow me,
> And I will bring thee where no shadow stays
> Thy coming . . ." (IV, 461–71)

This narcissistic moment, brought about by gazing with love and desire at a mirrored reflection, holds a certain amount of danger for Eve; thus the need of a warning from the unidentified voice.

The attraction of resemblances, the seductive illusion inherent in imitations and reflections, poses a threat to the proper understanding of truth, faith, and reality. In *Paradise Regained*, Christ warns that for men to peer with intensity at mirrored reality is to seek virtue in themselves, and to themselves arrogate all glory.[25] "Who therefore seeks in these / True wisdom, finds her not, or by delusion / Far worse, her false resemblance only meets / An empty cloud" (*Paradise Regained*, IV, 318–21). Milton here repeats the association made by Hobbes between resemblances and fanciful speculation, relating imitations to idle dreams and conjectures built upon "fancy" and nothing firm. Satan first begins his assault on Eve by whispering in her ear as she sleeps, "assaying by his devilish art to reach / The Organs of her Fancy, and with them forge / Illusions as he list, phantasms and dreams" (*Paradise Lost,* IV, 801–3). The next morning, Eve relates her dream to Adam, remarking on the vividness of the voice which she first mistook for that of her husband. The garden seemed more pleasant and fair to her fancy, as did the tree of knowledge; in her dreamlike imaginings, she becomes a goddess, leaving the confines of the earth and roaming with an unknown and disembodied companion into the air, ascending unto heaven and speculating on the life lived there by the gods and the angels. She is carried into the clouds and beholds the earth stretched beneath her, marvelling at the prospect wide and various, and wondering at her flight, change, and exaltation.

Adam likes not this uncouth dream of hers, and fears that some evil is the source and origin of this flight of fancy. He then offers an explanation of the workings of the fancy, as distinct from the understanding, that is remarkably similar to the definition provided by Hobbes:

> But know that in the soul
> Are many lesser faculties that serve
> Reason as chief; among these Fancy next
> Her office holds; of all external things,
> Which the five watchful senses represent,
> She forms imaginations, airy shapes,
> Which Reason, joining or disjoining, frames
> All what we affirm or what deny, and call
> Our knowledge or opinion; then retires
> Into her private cell when Nature rests.
> Oft, in her absence mimic Fancy wakes
> To imitate her; but, misjoining shapes,
> Wild work produces oft, and most in dreams,
> Ill matching words and deeds long past or late.
> (V, 100–114)

The fancy receives the impressions of things made upon the five senses, forming images and vague and indistinct shapes which reason (the understanding) joins together in utterances of affirmation and negation. This is called knowledge or opinion by Milton, and science by Hobbes.

When the reason is at rest—that is, sleeping—fancy often imitates the workings of reason, joining these "aery shapes" together in such a way that knowledge or opinion seems to be called forth. Such resemblances to truth and knowledge are often found in dreams; indeed, the vividness of dreams can actually make such illusions appear to be real and authentic. Yet, Adam warns Eve that the fancy can only match words and deeds in a vague and indistinct way, wildly and disjointedly with no firm relationship to the true and the real. This fancy, which seems to work most often in dreams, must be linked with the understanding; in tandem, they bring reason to the mind and to the soul. This reason is both intuitive and discursive, the former belonging more naturally to the angels, the latter more commonly to human beings. The implication is that discursive reason, that is, discourse, is often corrupt, pertaining as it does to the disputations and "quotations" of human life and activity. Discursive reason must be supported, and made significant and meaningful, by intuitive reason, which can be touched, immediately, by the Creator.[26]

I noted earlier that intuitive knowledge is related to those self-evident truths that are apparent to a mind touched and seasoned by God. In Book VIII, Raphael tells Adam to be lowly wise, and not question the ways of heaven. Adam should be concerned only with what concerns himself and his being on earth, and not to dream of other worlds or the creatures and conditions therein. Thus, true knowledge is based on moral rather than intellectual grounds; right reason is founded on obedience and subjection to God's just commands. Raphael's advice recalls the argument made in *The Reason of Church Government*:

> How much more aught the members of the church, under the gospel, seek to inform their understanding in the reason of that government which the church claims to have over them! Especially for that church hath in her immediate cure those inner parts and affections of the mind, where the seat of reason is having power to examine our spiritual knowledge, and to demand from us, in God's behalf, a service entirely reasonable. (*Reason of Church Government*, 440)

That such a service is entirely reasonable to Adam is apparent in his response to Raphael. He is satisfied that the prime wisdom is not to be discovered through a roving fancy and wandering thoughts; such curiosity breeds only anxious and molesting cares, vain notions, obscure and subtle thoughts that are remote and useless.

This intuitive reason, this lowly wisdom, proves to be problematic within the confines of Milton's argument in *Paradise Lost*. After all,

Milton's aim, by this great argument, is to assert eternal providence and justify the ways of God to men. Indeed, in *Areopagitica,* Milton argues that God encourages us in our "disputing, reasoning, reading, inventing, discoursing, even to a rarity and admiration, things not before discoursed or written of," and gives us minds that can wander beyond all limit and satiety.[27] Yet, despite this, Adam must confine his imaginings to the moral realm of ethics and behavior. Nature must be adapted to receiving useful information, and must be made "virtuous":

> The end, then, of learning is to repair the ruins of our first parents by regaining to know God aright, and out of that knowledge to love him, to imitate him, to be like him, as we may the nearest, by possessing our souls of true virtue, which, being united to the heavenly grace of faith, makes up the highest perfection.[28]

Adam must strive toward true virtue, understanding God "by orderly conning over the visible and inferior creature."[29] This virtue is comprehended by the reason only through an intervening act of God, through his illumination and guidance.[30] Likewise, by the exercise of Christian virtues, the reason can be made to apprehend and understand the prime wisdom itself.

By such an intuitive and subjective process, eternal truths of nature are well within man's grasp. Interestingly enough, this process takes place for Adam (if not for Eve) through a dreamlike state. At the end of Book XII, Adam tells Eve that "God is also in sleep, and dreams advise / Which he hath sent propitious, some great good / Presaging" (XII, 611–13). Earlier, he had related to Raphael that after God had instructed him about his eternal ways, a deep sleep had fallen upon Adam. In this state the cell of fancy, his internal sight, was left open, "by which / Abstract as in a trance" Adam saw the creation of Eve (VIII, 461–62). Earlier still, Adam was gently moved by an inward apparition created by his fancy, a vision that proved to be his first guide to paradise. In Adam's dream, the vision is abstract, not figured forth in any bodily image.

This kind of "sound," reasonable dreaming, is contrasted with that of the childlike, immature Eve. In Book IV of *Paradise Lost,* Eve, upon first awakening, hears the murmuring sound of water issuing from a nearby cave; the lake of water, spreading out before her, appears to her inexperienced eyes like a pure expanse of heaven, presenting a gleaming representation of all things contained in paradise. The shape she sees within, her own reflection, seems pleasing, sympathetic, all that is desirable. Only when a voice warns her that the shape is herself, and should not, therefore, be idolized, does Eve draw back from her image in the water, following that voice to Adam, "whose image thou art, him thou shalt enjoy / Inseparably thine" (IV, 472–73). In thus presenting Eve's progression beyond the narcissistic moment, Milton instructs the reader on the dangers of the unregenerate childhood state, when resemblances are often mistaken by the inexperienced and fanciful child for the things themselves, when playing with shadows offers greater rewards than understanding real objects.

Indeed, both *Paradise Lost* and *Paradise Regained* are poems about straight angles, about the inflexibility of the subject's unbending gaze and the potentially disastrous consequences entailed for that subject in looking awry at the creatures of the world. Yet, as Regina Schwartz rightly argues, although the poems deal with voyeurism and the scopophilic drive of the subjective gaze to dominate the object of desire, the poems cannot be read simply as an exercise in Freudian psychoanalysis.[31] The viewing subject, as much as the object, is always being watched by an invisible, inaccessible origin (the gaze of the Father) which is itself constantly under surveillance (or harassment, using Stanley Fish's terminology) by the reader. This tension is particularly notable in *Paradise Regained,* where even the Son, supposedly the mirrored image of God, becomes the object of scrutiny and desire on the part of the Father, Satan, the Narrator, and even the reader. This continuous displacement results in a serious intellectual disorientation and possible spiritual confusion on the part of any reading subject (the Narrator, the reader, Eve, Adam, Satan, and even the Godhead) acting in solitude and isolation. The problems of seeing, represented by the use of anamorphic similes and other devices subverting the mechanics of vision, are constantly foregrounded so that every reading subject ultimately realizes the futility of their autonomous will to knowledge, power, and dominance.

The complexities of vision, linked with the will to power, had previously been explored by Shakespeare who had created a complex register of metaphors, images, and other literary devices to demonstrate the matrices of perception and psyche. In the second act of *Richard II,* Sir John Bushy seeks to comfort Richard's Queen by describing her grief in terms of optical error. "Like perspectives, which rightly gaz'd upon / Show nothing but confusion; ey'd awry / Distinguish form; so your sweet Majesty, / Looking awry upon your lord's departure, / Find shapes of grief, more than himself, to wail, / Which, look'd on as it is, is nought but shadows / Of what it is not" (2.2.18–24).[32] Through this anamorphic simile, Bushy identifies and aligns optical illusion with fallacious understanding. Likewise, through this trope, Shakespeare questions how the mechanics of vision controls what the viewer, the audience, is capable of knowing and perceiving.

This metaphor of perspective, with its associative scientific terminology of the visual pyramid and the vanishing point, began to be used by writers in the seventeenth century as code words for discretion, judgment and cognition.[33] Galileo's observations with the telescope, and the experiments in optics by Descartes and Pascal, gave evidence that serious philosophical thinkers were consumed with questions about visual phenomena, especially when those queries sought to determine "truth" through a visual understanding of the world.[34] Scientific taxonomy, a primary method used to verify the order of nature, rested on several critical assumptions: the idea that nature was an orderly construct that could be known and described; that definition, based on taxonomy, constituted a statement of essence upon which scientific knowledge was built; and, finally, that words were "isomorphic with things."[35] Crucial here would be the faculty of vision to apprehend and then comprehend the essence of natural phenomena. Henceforth, the nature of the visible world would be controlled by the rules of optics.[36] While Italian artists had used linear perspective to express confidence in the certainty of all human knowledge, later thinkers and artists would challenge an unambiguous engagement and relationship between the knower and the knowable.[37] In contrast to most Renaissance painters, the artists of the seventeenth century were interested in a sophisticated use of lenses and optical devices. The telescope had opened up a new world to scientists and artists, so that the stable world of the Renaissance had been replaced by a more ambitious and problematic point of view. In the Baroque world, the viewer was not always separated from the painting, but would determine her own relationship with the artistic product. While scientists were consumed with notions of mathematical certainty and truth, Baroque artists were experimenting with a kind of art, notably anamorphic art based on distortion and illusion, that would be philosophically opposed to the kind of life-like representation of visible forms that were discussed earlier using classical notions of *mimesis*.

Indeed, in the sixteenth and seventeenth centuries, the use of anamorphic devices gave rise to a vicious and often slanderous series of treatises either denouncing the practice as deadly to the moral teachings of a Christian society, or as critically important to the need for skeptical doubt and uncertainty in both the arts and sciences. In Jurgis Baltrusaitis's book, it becomes clear that "anamorphosis renewed contact with the occult and at the same time with theories concerning the nature of doubt." The mirror took on magic powers by conjuring up phantoms.[38] For example, Henry Van Etten, in his 1633 work *Mathematical Recreations,* claims that "spectacles of Crystall cut with diverse Angles dimond wise doth make a marvelous multiplication of appearances."[39] These spectacles create delightful images, or the "delusion of nothing" (101). Anamorphosis shows the illusion in what should be the most accurate representation of reality, namely the rules and values of linear perspective. Instead anamorphosis, the strange parallax, demonstrates that a science can be used as an instrument or tool for producing dream structures, mock-realities, hallucinations, hysterical manifestations of truth, the darker grotesque shadows of the real world.

Jacques Lacan discusses specular anamorphoses at length in *The Four Fundamental Concepts of Psycho-Analysis.*[40] In this work, he highlights the inverted use of perspective in all structures of distorted or anamorphic representation. These specular devices, like all mirrors or glasses, imply not only narcissism but identification. For Lacan, this establishment of identity is founded on the idea of resemblances, imitations, the counterfeit, and the doppelganger, the reversal of the real. Anamorphic devices, according to Lacan, ultimately are a trap for the gaze, for anamorphosis plays havoc with elements and principles. Using the geometrical laws of linear perspective, anamorphic art does not simply and unambiguously render forms as they should appear, but projects them outside themselves and their formal space, distorting them so that they must be viewed from a certain point to return to normal. Reality and appearance are artfully and artificially separated by artists and poets. The end result is a remarkable distortion, all the more noteworthy because it is premised on—indeed, constructed by—the purely rational laws of geometrical linear perspective.

This fear of imaginative and fantastic illusion animates the work of not only Descartes but Pascal and the Port Royal logicians. The seventeenth-century preoccupation with anamorphosis, and the related concern with visual perception, addresses the multiple problems regarding the sign (the picture), the signified (the reality), and the connection between the two. For many seventeenth-century writers and thinkers, conceiving of thought as a mental picture had dangerous consequences.[41] According to Port Royalist Antoine Arnauld, since "we are able to conceive of a very great number of things of which we have no image at all," the imaginative faculty must be made subordinate to the pure intellect (*Logic,* 28). As such, it is false to suppose that all our ideas come through the senses, or have a real connection with any bodily image. We can express nothing by our words without having a clear and self-evident idea already present in the pure intellect without words or images derived from sense perception.

According to Lyons, "by locating the unique viewpoint in nontraditional ways—for example, far to one side and not on a line crossing the painted surface—the anamorphic artist reminds the viewer of the active role that the viewer must take in deciphering the image. . . . In this way anamorphoses . . . force on the viewer an attitude of obedience.[42] With anamorphic art, the beholder of the work must try to discover the correct viewing posture, thereby moving through several incomprehensible perspectives while determining that single, narrowly defined point of view. This procedure is meant to remind the viewer that in order to interpret any work of art an active role of deciphering must be assumed until the artist's intention is found. The artist makes that viewer aware of his or her own individual activity. Illusion and distortion are highlighted so that the viewer understands how easy it is to be distracted by visual seductions. The real and the true are thereby set in contrast to the apparent and artificial. The artist and the writer are presented in this scheme as the ultimate authorities that guarantee meaning and value.

In *Paradise Regained,* Satan wishes to place an anamorphic device between Christ and the panorama of the city of Rome, hoping to influence and change the perception of the austere and pious son of God. In the fourth book, Satan brings Christ to the western side of a great mountain, overlooking the plain of Latium and the city of Rome.[43] After surveying the seven small hills, the palaces, theaters, baths, trophies, and triumphal arches, indeed, all the ancillary adornments of worldly sophistication, the narrator informs the reader, and Christ, of a subtle and curious displacement in perception and discourse, interposed by a crafty Satan. "By what strange Parallax or Optic skill / Of vision multiplied through air, or glass / Of Telescope, [they] were curious to inquire," reads the text, the silent "they" implying both the reader and the as-yet-uninformed Christ (IV, 40–42). This displacement of an object, in this case, the city of Rome, due to a shift in the observer's point of view, is meant to trick the unwary reader into believing a falsehood to be a truth. The glass could be either a telescope or simply a system of mirrors meant to represent real things in the shape of a formless shadow.

Of course, the canny Son of God is not fooled by the illusions of art, calling this vision nothing else but dreams, conjectures, fancies, built on nothing firm, stabilized not by plain sense, but by fabling and smooth conceits. "Who therefore seeks in these / True wisdom, finds her not, or by delusion / Far worse, her false resemblance only meets," claims the Savior wisely (IV, 318–20). "He who receives / Light from above, from the fountain of light / No other doctrine needs, though granted true" (IV, 289–90). In other words, only by looking directly and rectilinearly at things themselves, illuminated by the pure light of reason, can any person hope to achieve wisdom, sapience, or any other kind of temporal or spiritual knowledge. In this instance, Christ denies even the idealizing potentiality of the mirror, that is, that traditional Platonic definition whereby the pure soul is a mirror reflecting the Good in the clarity of its surface. Instead, the speculum, for Christ and presumably for Milton, can only have an ambiguous position in the progression of knowledge, the image being necessarily inadequate. Even the clearest mirror, the mirror of Scripture, is ambivalent when compared to the clarity of that pure and intuitive intellect which does not rely on any physical or bodily forms. Christ has no need for intervening objects to instruct or seduce him; "and what he brings, what needs he elsewhere seek," says the son of God, noting that even books stand in the way of properly and clearly perceiving the divine. In *Paradise Regained,* the intervention of mirrors and writing, the objects of a making or a *techne,* serve only to distract and shadow the mind. The problem, of course, is that Christ himself is the embodied form of the eternal idea, the visible image of an invisible God, like the figure found in a mirror. As such, Christ himself, supposedly an example of the archetypal subject, is as much an object of the male subjective gaze as Eve herself.

Milton's consistent denial, yet subtle qualification, of the mirrored image is evident throughout *Paradise Lost,* most notably in the episode where Eve's narcissistic moment by the pool becomes an instance of anamorphic specularization, a parallax or displacement of gaze which must be corrected by a turning of the feminine I / eye to the masculine form of Adam. In the *Speculum of the Other Woman,* Luce Irigaray plays with the traditional idea of the speculum or mirror as an opposition to what she calls the dazzling fascination of the Sun—the image of the Good and the Beautiful.[44] Plato's analogy of the cave is the original speculum, "an inner space of reflection. Polished, and polishing, a fake offspring." The universe "is organized as mimesis; re-semblance is the law" (*Speculum,* 150–51). While discourse may still be divided along two separate poles, the serious pole (that is, the truth) and the playful or fictive one, the governing trope remains the mirror, rather than the light or the sun. Attached to the episteme of the mirror are the ancillary ideas of copies, fantasies, reflections, semblances, and specular anamorphoses, the parallax of Satan's trope. This parallax or optic skill could be a glass or a mirror or a telescope, or even a smooth conceit. In all forms, the speculum disperses and miniaturizes the potentiality of the enlightened and phallic gaze, reflecting only the painted figure, the fake, the seductive fantasy, the fiction. Irigaray's linguistic strategy subtly and strangely mimics that of the demonic monster, Satan. Her fakes and fictions, like those of Satan, are frozen by "likes" or "as ifs," smooth similes of evocation and figuration that interpose by some strange parallax or optic skill a fantasy that will weaken the potency of phallic perception. This views the mirror or speculum as essentially passive, "a snare of vain images that seduce us with a false vision of beauty and leave us with nothing."[45]

In the *Speculum,* Irigaray had demonstrated how Plato's episteme relied heavily on the idea of the mirror to shield mortal eyes from the blinding dazzle of the sun. "Reason," she writes, "which will also be called natural light—is the result of systems of mirrors that ensure a steady illumination (148). Being itself, whether we call it the Origin, the Idea, or, in metaphor, the sun, can never fully be confronted. Instead, the universe "is organized as mimesis; resemblance is the law" (150–51) Irigaray writes that only in rare moments of insight do we receive some kind of intuition that is prelinguistic and which will possibly match the Idea of the Good or the Beautiful. Augustine, for one, names this intuitive moment an intellectual apprehension which, while speech is being formed, hides itself in the secret recesses of the mind. For Irigaray, the intrusion of the speculum itself creates these hidden recesses, these cavities, spheres, sockets, chambers, and enclosures.

At the very center of Irigaray's *Speculum* lies a telling critique of

Descartes. For Descartes, man must trace for himself a true path, and not be swamped in a flood of dreams. She writes of him, "Since the ground threatens every minute to shake the present certainties of the subject, *it must not be allowed any power of specularization*. The basis for representation must be purged of all childish phantoms or fantasies or belief or approximations" (181). In order for this to happen, illusion must be constitutive of thinking, illusion serving "as fiction of proof of the cogitatum itself" (182). The other, that is, the illusion, becomes the mirror in which the *I* is reflected. Irigaray goes on to equate women with this illusion which makes thinking possible. This exercise in Cartesian thinking demonstrates how women are thereby denied any kind of independent visibility or reality, and are instead specularized and made invisible using the laws of anamorphosis. As such, the trope of anamorphic specularization mimics the deceitfulness, and therefore marginality, of not only the feminine gaze, but the visible reality of woman. In order for the woman to view her own distorted form with any kind of intelligibility, she must leave her initial position and assume a predetermined and narrowly defined viewpoint.

In Book IV, Eve recounts to Adam how her first "unexperienced thought" came about through gazing at her reflection in the lake. Eve is in need of a warning from a wiser, more experienced, though disembodied and unidentified, being; as such, this moment of mimetic and narcissistic love and desire brings spiritual danger to her. Jacques Lacan identifies this moment in a child's development as the *stade du miroir*, or the mirror stage.[46] Between the ages of six and eighteen months, the child, when first seeing a reflection in a mirror, takes that reflection to be real. Later, at the moment of self-recognition when the child learns that the image is a reflection of himself, the child will become captivated with that image, seizing hold of the mirror, playing with it, posing before it. The mirrored image, bringing forth as it does such feelings of joy, pleasure, and jubilation, holds a unique and powerful place in this stage of development. Lacan accounts for this by noting that since the child at this age is almost totally dependent on the mother, the mirrored image offers an alternative, integrated being which the child can control, providing compensation for the insufficiency of self that the child possesses and leading subsequently to an imaginative construction of the self that child will become. Lacan understands this mirror stage as an identification, a "transformation that takes place in the subject when he assumes an image . . . the symbolic matrix in which the *I* is precipitated in a primordial form, before it is objectified in the dialectic of identification with the other, and before language restores to it, in the universal, its function as subject."[47]

This definition of the mirror stage effectively illuminates that moment by the pool when Eve sees her reflection and pines with vain desire. The voice calling her away from this self-absorption of a shadow brings her attention, her gaze, to the image of another, that is, Adam. The call beckons Eve to a more adult stage of consciousness and knowledge, where objects and their boundaries are more firmly defined and perceived. The infant, newly born, has yet to establish parameters for her world. The infant world and God's word are intimately linked. The alternative and metaphoric world, though resembling the natural world, is inhabited by an other who obviously can communicate through other means besides speech and language, who can rely upon other senses besides the hearing faculty. As we have seen, it is just this kind of speechlessness which is a threat to Eve in Milton's paradise. She can become a fully adult woman only when she can communicate with Adam, thereby taking her place in the adult world of speech and grammar. The stage of primary narcissism must yield to a secondary (and perhaps less satisfying) narcissism which entails an identification with the other, locating "the imaginary and libidinal relation to the world in general."[48] In this episode, Adam claims Eve as his flesh, his bone, his soul, and his other self. In response, Eve defines herself, conceptually and linguistically, in contrast to Adam. "I yielded, and from that time see / How beauty is excell'd by manly grace / And wisdom, which alone is truly fair" (IV, 489–91). Not only does Eve learn to speak, she also learns to divide and define reality along separate and opposing registers, feminine beauty opposed to manly grace and wisdom. Her creation continuously refers back to Adam, for him she was made and from him. Without Adam, Eve has no identity, the "I am" leads inevitably to no end. The assumption, in the Cartesian sense, is not *cogito ergo sum*, but *cogitat Adam, ergo sum* (Adam thinks, therefore I am).

The oppositional registers are operative throughout the poem. The newly born Eve, unlike Adam, remains initially "reposed," lying back beneath a shade. (The Latin also implies replacement, restoration, substitution.) She is reclining, passively and modestly covered by a shade (Adam, upon first awakening, finds himself lying in the sun, whose beams quickly dried the balmy sweat from off his limbs,

VIII, 254–55). She immediately wonders (thinks) what she was, where she came from, and how. Upon her first awakening, she *hears* the murmuring sound of water issuing from a nearby cave. Adam, in contrast, *sees* the "Hill, Dale, and shady Woods, and sunny Plains, / And liquid Lapse of murmuring Streams" (VIII, 262–63). The word *lapse* in this passage implies a slip or fall, into an inferior position, perhaps, which is strongly tied to the sound, and presumably the image, of the murmuring stream. In Book IV, this murmuring sound of waters promises to be a similar snare for Eve. The lake of water, spreading out before her, appears to her inexperienced eyes like a pure expanse of heaven, presenting a gleaming representation of all things contained in paradise. The lake is immediately identified as a mirror, "its surface unmoved." Presumably, in the episode in question, there is no apparent distortion in this flat mirror, its smooth surface presenting a fiction of reality rather than a perversion of it. Upon gazing, or embracing, the surface of the mirror, Eve soon sees a moving figure. The shape she sees within, her own reflection, seems pleasing, sympathetic and all that is desirable. With her lack of experience, Eve mistakes this wonder for another sky.

As we have seen, the mirror image has always had a supernatural quality; mirrors were often used as instruments for conjuring up ghosts and the dead. Yet, the mirror also has been used to demonstrate uncertainty and limits. One can use a well-placed mirror to see what goes on behind our backs, for example. A combination of mirrors, convex, concave, or otherwise, can also be used to demonstrate often otherwise invisible phenomena. In Van Eyck's *Arnolfini Marriage,* a cleverly placed mirror allows the viewer to see the wedding guests, and, if tradition can be believed, the artist himself. In the seventeenth century, Velasquez tried a similar trick; yet, rather than using a convex mirror, Velasquez uses a flat mirror, and thereby refrains from playing with any of the usual laws of perspective. According to Lucien Dällenbach, in his book *The Mirror in the Text,* by the use of a flat rather than a convex mirror, "Velasquez' painting achieves a reciprocity of contemplation that creates an oscillation between the interior and the exterior, making the image 'come out of the frame,' while inviting the visitors to enter the picture."[49]

This distinction is important for this study of Eve. That is to say, the distortion does not lie in the materiality of the flat mirror, but in Eve's perception of her image, the essential nature of her gaze. Eve is given a kind of double plot or world which she must choose from. Of course, the mirror world is much more enticing and engaging. She sees a moving figure bending down opposite her. The mirror sensitively responds to every move she makes, duplicating her action and reflecting it in miniature. This divided unity naturally enchants the isolated and solitary Eve. However, she still lacks the self-awareness which would allow her to see this image as divided rather than single. As it stands, she mistakes the image for the reality. After her initial fright, however, Eve returns, pleased, answering the reflective gaze with sympathy and love. Curiosity, the eagerness to discover, is linked with affection and perhaps even physical desire, the libidinal instinct. After several attempts at further communication, Eve pines with vain desire, for the image, the fantasy, the reflection, can never return any favor.

Only when a voice warns her that the shape is herself, and should not, therefore, be idolized, does Eve draw back from her image in the water, following that voice to Adam. The implication here is that her gaze has perceived (or conceived) but a shadow, not something real or meaningful, in a sense, creating a distortion, a parallax or an anamorphic device. In order to give meaning to that distortion, Eve must find a proper viewpoint or reference point. Another feature of anamorphic art is that it not only invites the viewer to contemplate the work from a particular viewpoint, but punishes the viewer if she does not do so, "by being confronted by meaningless form."[50] As such, Eve shifts her gaze and stares at Adam, "whose image thou art, him thou shalt enjoy / Inseparably thine" (IV, 472–73). Strangely, that voice does not suggest that Eve turn her eyes to the sky, towards Heaven, thereby correcting her mistake. It is Adam's image that she must "enjoy." To look at Adam is to see her self interpreted, to see a meaning in another's face. Presumably, Eve is the mirrored image of Adam; when she looks at herself, she must instead see him. Also, when Adam gazes upon her, he must see himself. As Irigaray suggests, Eve is denied her own narcissistic moment: "woman will not choose, or desire, an 'object' of love but will arrange matters so that a 'subject' takes her as his 'object'" (113). The instincts of the inexperienced female are transformed. Yet this transformation of the gaze does not lead her to a greater understanding of things themselves, that is, the heavens; instead, her gaze is refracted and displaced by the authorized perception and discourse of Adam, whose purpose is correctly to interpret the dreams and fancies of his feminine helpmate.

Indeed, in Book IV, Eve is presented by Milton as a naive and inexperienced reader, dazzled by her own reflection. The reader has already seen how a narcissistic libido caused Eve to make her first mistake in identifying a proper object of desire. By pulling Eve away from her own reflection, and allowing that gaze to be refracted through Adam, Milton cautions the reader about an unregenerate state of being that is childlike and feminine, where fake objects are loved and desired as much as originals, and where the inexperienced and fanciful woman (or child) is seduced by illusions and distortions. One remembers Irigaray's rendering of Descartes in *Speculum of the Other Woman*. For Descartes, writes Irigaray, "the basis for representation must be purged of all childish phantoms or fantasies or belief or approximations" (181). It becomes necessary to find one fixed point, in order to distinguish the real object from the virtual object, the confusion of which, according to Irigaray / Descartes, "may persist even in one who has faced up to the laws of optics" (189). That is to say, anamorphic displacement both affirms and denies the geometrical law of optics, saving both the appearances and the skeptical doubt of all surface illusion. In Milton's poem, Eve's image on the surface of the pool is both affirmed and displaced by Adam's own narcissistic and libidinal instincts. In fact, Adam, unlike Eve, comes equipped with a faculty that allows him to negotiate the space between the real and the copy, a faculty which succeeds in interpreting the various images or shadows found in dreams and in mirrors.

Indeed, while Eve, upon first seeing her husband, denies his appropriation of her image, thinking him "less winning soft, less amiably mild, / Than that smooth wat'ry image," Adam immediately claims her as his other half, created from his flesh and his bone. Her visible reality is affirmed by the enamored gaze of Adam ("And in her looks . . . inspir'd / The spirit of love and amorous delight," VIII, 474–77). He becomes the glass through which the correct posture, or reading position, can be determined for Eve. The speculum for Eve becomes not the mirrored surface of the pool, but the perception and the discourse of her male counterpart. In God's words, she is Adam's "likeness, (his) fit help, (his) other self, / (His) wish, exactly to (his) heart's desire" (VIII, 450–51). For Adam, the figure of Eve simply disappears, for when he gazes upon her, he sees the image and resemblance of himself. "I now see / Bone of my Bone, Flesh of my Flesh, my Self / Before me" (VIII, 494–96). Unlike Eve, Adam is allowed to turn (or displace) his own gaze upon many different kinds of natural phenomena, and still see himself (and his God) reflected there.

In the temptation episode of Book IX, an alternative, and more sinister, endorsement of mirrored images is used. Instead of Raphael, the teacher of Eve is Satan; he uses her inexperience and her vanity against her by playing on the idea that humans were made in the image of God. He reminds Eve that numberless angels wait upon her train daily and that she is adored and served as a goddess among gods, effectively turning her into an idol of herself. At the same time, Satan encourages her to revel in the pleasure that she finds in herself, and to discover in her mirrored reflection the image of the Creator. There is a noble element in this advice from Satan, just as there was a sense of joy in the dream that he had created for her earlier. Yet, as in all of Satan's endeavors, any worthy aspect is corrupted by his emphasis on self above all else.

In Eve's case, as in Satan's, the preoccupations with dreams and the self, whether real or found in a mirrored reflection, are idle pursuits. Alternatively, with Adam, eternal truths about the world and the self can take place in dreams. Several times in the course of the poem, Adam is brought to self-awareness through dreams and visions. In Book XII, Adam tells Eve that "God is also in sleep" and that dreams advise and send propitious and good tidings. Earlier, Adam told Raphael that he had fallen into a deep sleep after God's instructions, and therein witnessed the birth of Eve. Earlier still, the cell of fancy, his internal sight, had been left open, creating a vision that had proved to be his first guide to paradise. Of course, this initial dream was presented in an abstract way, not figured forth in any bodily image. The soul, the mind, should progress to that state where custom, linguistic or otherwise, no longer interferes with the communication with God. Thoughts, then, would seem to take precedence over things, at least for Adam if not for Eve. Here the path towards identification with the godhead must necessarily begin and end through the speculum of the male gaze which transcends the materiality of things in order to grasp the intelligibility of pure thought.

Despite this transcendence of the linguistic, visual, and sensual, Adam is presented in the narrative as transformed by gazing upon the visible sign of Eve. In Book VIII, Adam first sees Eve in a dream,

## CHAPTER 6: THE FIGURE IN THE POOL

fashioned from his rib and appearing manlike but of different sex. At her disappearance, Adam is left in the dark, deploring his loss, the absence of the feminine. Later still, after naming her "Woman, of Man extracted," he waits for her obsequious majesty to approve his "pleaded reason" (VIII, 510). At this moment, which corresponds to Eve's sexual yielding in Book IV, Adam, who had until this moment spoken primarily in abstract language, takes his place in the feminine world of nature and poetic symbol. Word and thing is harmonized in the narration of nuptial bliss. "All Heav'n, . . . Gave sign of gratulation, and each Hill; / Joyous the Birds; fresh Gales and Gentle Airs / Whisper'd it to the Woods, and from thir wings / Flung Rose, flung Odors from the spicy Shrub" (VIII, 511–17). The fusion of masculine and feminine, symbolized by the sexual act (Adam's "pleaded reason"), allows for the creation of poetic order. This linguistic and conceptual transformation is a result to Adam finding his own identity by gazing upon the inverted "masculinity" of Eve ("manlike, but of different sex").

Ironically, it is necessary for Adam's education to see himself in Eve, noting the inversion, the difference, the otherness. Again, a reference to the world of painting and picture-making is in order. Alberti writes that "the tale of Narcissus fits our purpose perfectly. What is painting but the act of embracing by means of art the surface of the pool?"[51] Likewise, Milton tells his reader that a new system of perspective must be discovered which will more adequately govern our visual and mental perceptions. By looking into a mirrored image that is identical to the natural world, yet which constantly reveals its otherness and difference as an artifact or counterfeit, the regenerated eye will successfully dismantle artificial barriers and more clearly understand the world, *and a bit more*. Mental cogitation, asking "what can it mean," negotiates that space between the material world and the world beneath the watery glass, or in dreams. As such, the mind should better regard the surface of the pool, the reversed image, and not be content to play with (to understand) the toys and baubles of the material world. To disregard that reversed reflection in the pool is to confine oneself to the vain and deceptive devices of this world. To reverse the usual order of things, a new perspective must be found.

Adam, with this new instrument (an inverted Adam named Eve) will see (and conceive) his own felicity, glory, and bliss. What is emphasized is the quality of perception, linked with desire, that illuminates and levels the natural world, its objects, to a meaningful and comprehensive location. Transparent words—transmitting rays of light without diffusion so that bodies can be distinctly seen without distortion—allow natural objects to shine through. Words should not distort these natural objects but must be a visible transmitter, a mediator, a polished glass without any warps or unnatural bents. This new speculum of man, which resembles the speculum of woman described by Irigaray, is allowed an extra dimension or purpose; the mirror, then, has a dual and hence ambivalent function for both Adam and Eve. Anamorphosis, governed not only by rational and mathematical laws but by religious and quasi-mystical intuitions, reminds the viewer that the world of appearances is deceptive. Often a mirror, a speculum, or a screen, must be used to determine a form and a meaning that is itself constructed, ironically, upon rational principles. The viewer's role as interpreter and reader is extremely self-conscious and active, actively negotiating meaning with the agent of origin, be that God or the poet. Yet, the reader also is encouraged to be intensely individualistic, lonely, and isolated. Meaning, though entirely dependent upon the viewer, is controlled completely by the rules of the discipline. The viewer must be instructed by these rules in order to find the proper, and singular, point of reference. For Milton, the parallax is a convenient trope, for this kind of optic skill reminds the viewer that all is not what it appears, that the physical image before the eyes is a sham. The remedy, for Milton, for Christ, and for Adam, lies in focusing the eye beyond the materiality of the image, and gazing instead upon the workings of the mind that apprehends, judges, and even creates that image, distorted or otherwise. He is thereby separated from the realm of material and physical forms, in the hope of removing himself from the flawed world around him. The reader is alone now, caught in their own theatre of transcendent intelligibility. As Irigaray claims, "everything foreign, other, outside its present certainties no longer appears to the (masculine) gaze . . . (he is) no longer able to make out, imagine, feel, what is going on behind the screen of those/his ideal projections, divine knowledge" (362).

We can ask, then, what this system of reversed images implies for Eve, or the female reader. This specular system of mirrors, while undoubtedly functioning well for Adam, is extended to Eve, who must continually fix her gaze upon Adam in his physical (and inverted) form. The unconditional certainties of the self-conscious thinking mind, the *cogito sum,* is thereby thrown in doubt by the difference of Eve. The *vanitas* demonstrated to Adam by the anamorphic trope succeeds in questioning the power of the masculine mind. From the speculum of woman, from the female perspective, this theatre of intelligibility remains a cave of likenesses and resemblances; the speculum of the Father, reflecting the image of the son, is not a source of singular identity and origin, but the beginning of difference, and even rivalry. Perhaps, in the end, this is the meaning of Eve's, and Adam's, speculative gazes in *Paradise Lost,* to establish a relationship of combative difference between the philosophers and the poets, the imprisoned children of the cave.

# Epilogue

> ... and there is no new thing under the sun.
> —Ecclesiastes 1:9

I BEGAN THIS STUDY BY BEGGING THE QUESTION NECESSARILY posed by any *episteme* which consistently challenges philosophy with rhetoric, the serious with the playful, the fictional with the true. My central concern asks why metaphorical playfulness, the rhetorical ideal of life, can never be serious; why words (*verba*) are constantly deemed inferior to ideas (*res*). The Western sensibility, trained on Plato's opposition to sophists and poets, continually skews any complementary relationship between *res* and *verba,* usually to the detriment of words rather than things. In the present age, this opposition is still figured in the battle between the arts and the sciences.

More recently, literary theorists have entered the fray, adding a scientific dimension to the study of metaphor, poetry and fiction. The discursive and critical enterprise, the metalanguage of criticism, has come to replace in value and priority the original literary (and figurative) work of art. The study of fiction has become deadly serious. My aim in this study thus started with a question concerning the schism of words and things in the seventeenth century, and came to include an awareness of similar issues raised by modern literary theory. Part of my concern was to place a seventeenth-century issue in relation to the current state of literary and critical practice. My own work (and practice) tends finally to accentuate the differences, to locate texts firmly within their own historical context and to find within any critical stance a solution to the writers' contemporary concerns.

Any linguistic epistemology is necessarily tied to the given historical situation. Such an ideology is a necessary fiction, a model or a paradigm with which, and through which, a society is able to understand itself and its behavior. Yet, this model, in order to be effective and to sustain societal order and control, must be believed by the majority who think and believe that this model is "true," that meaning is authentic. Herein lies the problem. How to convince the populace that the fiction, the ideology, can be guaranteed by an objective outside authority. In a study such as this one, the critic asks how and why a culture subliminally adheres to the terms of the contract inherent in any ideology. Debora Shuger calls these interpretive categories the "habits of thought" expressed by the dominant players of any culture.[1] Such a critic will question the decisions that are made by any given culture to enable them to live in their world, their society. Yet, even asking this kind of question places the critic within the late twentieth century, as a member of a literary culture which constantly tries to cope with a hostile environment.

The preceding work has attempted to make a contribution to that field of literary criticism which unites the study of language theory with an awareness of the particular historical situation. With this in mind, I choose to end the work with a traditional peroration which, like Folly's, recalls and revises much of what came before. That is to say, rather than locating Bacon's work exclusively in his own historical moment, I would like now to clarify the similarities between Bacon's epistemological stance and that of recent intellectual historians and literary theorists. Modern theorists, following the lead of Foucault, denigrate Bacon as the chief architect of an *episteme* of identity which strangles and restrains the free interplay of signifiers in favor of a strict bilateral identification of word and thing, sign and signified. The final fissure between words and things is generally considered to take place in the seventeenth century, with the scientific rationalism introduced by Francis Bacon and perfected by Thomas Hobbes and the Royal Society.

Rosalie Colie sees Bacon as an orderly and rational thinker who hopes to make all things "clear" for the reader, all causal connections so self-evident that the scientist can never be caught cheating or lying, and calling for the truthful, the exact, the definite restatement of things in univocal rather than equivocal terms.[2] Likewise, Wolfgang Iser would place Bacon within the group of critics who emphasize the explanatory function of literature, finding "psychological satisfaction.... in creating certainty where 'the nature of things doth deny it.'"[3] Yet, much of my study has shown that in Bacon's rejection of peripatetic rationalism, and his subsequent advocacy of pre-Socratic reasoning in terms of broken and inexact knowledge, Bacon departs from this kind of epistemology based on certitude. The discursive enterprise, while important, is necessarily subordinate to the intuitive insight that precedes linguistic utterance. Bacon is as enthralled within an episteme of resemblance as any deconstructionist critic.

Certainly, myth can and should be explained and clarified, just as metaphor can be redescribed in literal and prosaic terms. Yet, the new instrument of science, the *novum organon,* relies upon a faculty which is able to see the similar within the dissimilar, to create scientific metaphors that serve as heuristic models of discovery. The scientist, then, should take as a role model the makers of myth and fiction who hide the connection between word and thing under a mist of ineffability. As Augustine claims, knowledge floods the mind with a sudden flash of light which is often gone in an instant. The expression of this knowledge comes after, when the "intellectual apprehension has already hidden itself in (the mind's) secret recesses."[4] It is almost impossible to match this intuitive knowledge with accurate, isomorphic words. It is appropriate to highlight this linguistic impossibility through some kind of representative illusion, which, Ricoeur tells us, claims the impossibility of "uniting the interiority of a mental image in the mind and the exteriority of something real that would govern from outside the play of the mental scene within a single entity or 'representation.'"[5]

Simple observation of nature leads, for Bacon, to that intuitive moment when axioms and experiments are perceived and conceived by the intellect, through an apprehension of similarities and dissimilarities. Boyle calls this intuitive faculty of apprehension a "clear light" by which the "inquisitive and well-instructed considerer" might well understand those truths of nature that God has left in the physical world.

> God has couch'd so many things in his Visible Works, that the clearer Light a Man has, the more he may discover of their Unobvious Exquisiteness, and the more clearly and distinctly he may discern those Qualities that lye more Obvious. (Boyle, *Christian Virtuoso,* 15)

The emphasis is not upon either the sign or the signified, but upon the perceiving eye which plays between the two, considering not only the obvious and clear meanings, but the ambiguities as well. Nature, and fiction, is not read for its explanatory function, but for its impact on "the inquiring mind that knows how to probe the ambiguities of vision."[6]

The fundamental dilemma faced by seventeenth-century writers concerning the problem of writing and reading, namely how to invent an appropriate style for the transmission of newly discovered scientific and philosophic matter, has yet to be resolved. The recent critical theories of language pondered and articulated by Derrida, de Man, Ricoeur, to name but a few, testifies how ancient problems can be articulated in modern dress. Literary theory continues to question how "meaning" is created in language, and how the meaning might be transmitted without losing either the integrity of the idea, or its complexity. Jacques Derrida asks that the linguist/philosopher not find this meaning in either the signified (the thing represented) or the signifier (the semantic unit representing that thing). Instead, meaning comes from the mind's active play between the two, sign and signified, word and thing, style and matter. The resulting discourse, the meta-language that results, becomes the "true" yet still provisional meaning of any text.

A similar kind of approach is found in Bacon. The reader is asked in a true induction to gather the evidence, the details of words (style) and things (matter), to reassemble them in light of perceived similarities or dissimilarities (identity or *différance*) and finally to determine axioms that will lead to further experimentation. The discourse that results from this method (the experiments reduced to descriptive narrative) does not replace the text, but serves as a supplementary and provisional substitute, until the next scientist/critic probes the inner recesses of the material. This delving into the labyrinth has its roots in the type of scepticism which anchored Bacon's "restauration" of knowledge so firmly in nature. Closure of meaning is emphatically resisted, unless it be centred within the mind's play between sign and signified. "And lastly, aphorism, representing a knowledge broken, do invite men to inquire further; whereas methods, carrying the show of a total, do secure men, as if they were at furthest" (*ADV,* 136). Meaning is thus not located strictly within the referential claim for truth, within things or the signified. Neither is it found in the logical structure of the signifying text. The rationalism found in peripatetic science is deemed inadequate, as is the notion that logical coherence alone can be a criterion of scientific truth. In this, Bacon roots his empiricism in the thinking of the pre-Socratic philosophers of nature, who devised a "poetics-of-things" by which metaphors constitute the text of the world.

Yet, in Bacon's scheme the "essay" which explains the myth is not taken to be the primary vehicle of truth, as it is in the work of Derrida or de Man. In prefacing de Man's *Blindness and Insight*, Wlad Godzich posits a difference between aesthetic language and the "language of inquiry governed by reason."[7] This latter phenomenon is likened to a flash of lightning which reveals the "inner configuration of the surrounding landscape and the forces at play within it. . . . The flash is not the secret but the occasion of the moment when all is in the light; the reward for peering into the dark."[8] In this sense the critical discourse, trained upon reason and logic, becomes the primary discourse, allowing the "landscape" to speak in its own voice, and no longer through figurative expression. The discursive and critical practice becomes superior to the intuitive one, for the latter, while gaining access to truth, leaves it ineffable. In contrast, "the object of a methodological practice is to give truth or meaning a vehicle in which it can make itself manifest unmediated," speaking in its natural, rather than its artificial voice.

One suspects that Bacon himself would call such reasoning an Idol of the Mind as well as the Marketplace. As I noted before, rhetoric itself is a constructed concept, an artificial animal that can serve various functions. Nature can never speak in its natural voice, as Godzich claims for de Man's meta-language, for any utterance, even the most transparent or opaque, is in *its* nature contrived and stylized. Bacon recognizes this, and in his *Instauratio Magna* relies on an *experientia literata*, rhetorical in nature, which precedes the new organon, or *interpretatio naturae*. His new method employs the scientific reader's intuitive faculty that can both create and decipher figurative language in the process of further discovery. The mystery and illusion inherent in myth and fable is used by Bacon not only to mimic nature in an entertaining performance, but to demonstrate that obscure and broken knowledge, discourse clouded in a dense mist, is the most appropriate avenue of discovery.

These speculations on fictional and artificial discourses as a way of discovering the natural brings me full circle to the place where I started, namely to current research on what Paul Rabinow calls biosociality in the impulse behind the creation of artificial natures, that is, the human genome project. Like Rabinow, my own position on this issue is one of neither commitment nor opposition. The possibility of opposition would no doubt be linked to traditional, and unconscious, ideals of a residual naturalism, derived ultimately from the Greeks, which holds that "the artificial is never as good as the natural; generation furnishes the proof of life (life is autoproduction); homeostasis (autoregulation) is the golden rule."[9] Yet, the French philosopher François Dagognet claims that nature has not been natural, in the sense of distanced from human works, for millennia. As quoted by Rabinow, Dagognet "asserts that nature's malleability demonstrates an 'invitation' to the artificial. Nature is a blind bricoleur, an elementary logic of combinations, yielding an infinity of potential differences.... If the word 'nature' is to retain a meaning, it must signify an uninhibited polyphenomenality of display."[10] We can almost hear the echo of Bacon's phrase *ipsissimae res* ("truth and utility are the very things themselves" aphorism 124). Dagognet (like Bacon) offers the following claim: "Either one adopts a sort of veneration before the immensity of 'that which is' or one accepts the possibility of manipulation."[11] Facing the complexities and ambiguities inherent in this position continues to challenge our understanding of a Baconian construction of nature and technology.

# Notes

## Prologue

1. Max Horkheimer and Theodor W. Adorno, *The Dialectic of Enlightenment,* trans. John Cumming (New York: Continuum Publishing Company, 1993).
2. Jürgen Habermas, "The Scientiation of Politics and Public Opinion," in *Toward a Rational Society: Student Protest, Science, and Politics* (Boston: Beacon Press, 1970, 63–64).
3. Carolyn Merchant, *The Death of Nature: Women, Ecology, and the Scientific Revolution* (San Francisco: Harper & Row, 1980), 33. Merchant also distinguishes Bacon, who advocated a torturing and altering of nature, from Valentin Andrea, who suggested that nature should be observed and emulated ("aped"). Part of my task in this book is to argue that Bacon's stand on nature is not as distant from Andrea's as Merchant supposes, that both are responding to common patterns of thought whereby (the book of) nature was thought to fashion and change the observer.
4. Paige Dubois, "Subjected Bodies, Science, and the State: Francis Bacon, Torturer," in *Body Politics: Disease, Desire, and the Family,* ed. Michael Ryan and Avery Gordon (Boulder, CO: Westview Press, 1994), 186.
5. Antonio Pérez-Ramos claims that, "to contemporary philosophers and ensuing generations, Bacon was the author of *Novum Organon* (1620) and *De Augmentis Scientiarum* (1621), was semi-Paracelsian, semi-animistic, geocentric" (7). He was also an embarrassment to the Corpuscularian and Copernican Baconians of the age. *Francis Bacon's Ideas of Science and the Maker's Knowledge Tradition* (Oxford: Clarendon Press, 1988).
6. Dubois, "Subjected bodies," 187.
7. Diderot and D'Alembert admired Bacon's naturalism, experimentation and positive evaluation of the mechanical arts. For them, "science was perceived as the hallowed way of dispelling superstition and ignorance and, therefore, as the best means of attacking the political and social system which purportedly rested on them: royal absolutism, Church authority, and the feudal order"; Antonio Pérez-Ramos, *Francis Bacon's Idea of Science and the maker's Knowledge Tradition,* 20.
8. Pérez-Ramos, *Francis Bacon's Idea of Science and the Maker's Knowledge Tradition,* 48. I also use Pérez-Ramos's method of identifying "ideas" of science at "different levels of depth and awareness, in a variety of fields, be it philosophical, scientific or literary" (48n1), as opposed to "ideals" of science. The latter indicates "how a scientific community imagines science as it ought to be if ever completed" (48n1) and expresses a criterion of rationality (about measurement, procedures, assumptions, and so on), by which ideas are either accepted or rejected in the name of that ideal.

See also A. Funkenstein's *Theology and the Scientific Imagination* (Princeton: Princeton University Press, 1986), 18–22.

9. According to Pérez-Ramos, the "meaning of *sophia* and *epist{e}m{e}* and *téchn{e}* were practically indistinguishable in the fifty century and the Sophistic age" (*Bacon's Idea,* 55n9).

10. See Richard A. Lanham, *The Motives of Eloquence: Literary Rhetoric in the Renaissance* (New Haven: Yale University Press, 1976), 6; and Stanley E. Fish, *Self-Consuming Artifacts: The Experience of Seventeenth-Century Literature* (Berkeley: University of California Press, 1972), 24.

11. See Barbara Shapiro, *Probability and Certainty in Seventeenth-Century England: A Study of the Relationships Between Natural Science, Religion, History, Law, and Literature* (Princeton: Princeton University Press, 1983). Problems relating to rhetoric are central to Shapiro's thesis, which argues that the breached epistemological barrier between logic and rhetoric, knowledge or opinion, made possible the distinct intellectual style of nearly all seventeenth-century English writers and thinkers.

12. This Baconian epistemology has, of course, attracted the attention of modern scholars. Michel Foucault distinguishes between a Renaissance "episteme of resemblance" whereby the sign implies three distinct elements, the sign, the signified, and that which makes it possible to see in the first the mark of the second, and a neo-Classical "episteme of identity," a strictly binary organization of sign and referent. *The Order of Things* (London: Tavistock Publications, 1970), 64. Timothy Reiss classifies the two prominent discursive practices as "patterning" and "analytico-referential" discourse, in *The Discourse of Modernism* (Ithaca: Cornell University Press, 1982), 30. While both associate Bacon with the referential episteme, which assumes that the external world is a fixed object separate from the discursive forms used to represent it, my study places Bacon firmly within the "episteme of resemblance" and "patterning."

13. See Paul Ricoeur, *The Rule of Metaphor,* trans. Robert Czerny (London: Routledge and Kegan Paul, 1978), 39. Sir Philip Sidney likewise encourages an apt "feigning" of notable images, virtues and vices, in the service of delightful teaching, so that the end of all earthly learning is virtuous action; *A Defence of Poetry,* ed. Jan Van Dorsten (Oxford: Oxford University Press, 1966). For an elaboration of the paradoxes inherent in *mimesis,* and the notion of originality in the counterfeit, see chapter 3.

14. Paul Rabinow, "Artificiality and Enlightenment: From Sociobiology to Biosociality," in *Incorporations,* ed. Jonathan Crary and Sanford Kwinter (New York: Zone Press, 1992), 148.

15. Rabinow, 149.

16. Donna Haraway, "A Manifesto for Cyborgs," *Socialist Review* 80 (March–April 1985), 100.

17. Ibid., 70, 67.

18. Pierre Gassendi, *Syntagma Philosophicum,* in *Opera Omnia Book I: De logicae fine* (Lyons, 1658), 65.

## 1. Renaissance *Res* and *Verba:* Toward a Poetics of Truth

1. Sir Philip Sidney, *A Defence of Poetry,* 23.

2. Ibid., 25.

3. For this debate on rhetoric and philosophy, see George A. Kennedy, *Classical Rhetoric and Its Christian and Secular Tradition* (Chapel Hill: University of North Carolina Press, 1980), Charles Sears Baldwin, *Medieval Rhetoric and Poetic* (New York: The

Macmillan Company, 1928), Samuel Ijsseling, *Rhetoric and Philosophy in Conflict* (The Hague: M. Nijhoff, 1976), Paul Oskar Kristeller, *Renaissance Thought and Its Sources* (New York: Columbia University Press, 1979). The quotation from Gombrich is taken from his book, *Art and Illusion: A Study in the Psychology of Pictorial Representation* (Princeton: Princeton University Press, 1969), 395.

4. See Terry Eagleton, *Literary Theory: An Introduction* (Oxford: Basil Blackwell, 1983). Eagleton refers to this mode as the classical "contractual" mode of language, where words are simply tokens of exchange allowing for the transaction of prelinguistic experiences. In this scheme, meaning is a commodity possessed by each individual (115).

5. Howell. "*Res et Verba:* Words and Things," *English Literary History* 13 (1946): 131.

6. Quintilian, *Institutio Oratio,* Book 8, Preface, 20, 21, trans. H. E. Butler. Cambridge, MA: Loeb Classical Library, 1958. "Curam ergo verborum, rerum volo esse sollicitudinem. Nam plerumque optima rebus cohaerent et cernuntur suo lumine."

7. Hobbs, *Leviathan,* ed. John Plamentz (Glasgow: William Collins Sons & Co., 1962), 85.

8. Else, "'Imitation' in the Fifth Century," *Classical Philology* 53 (1958) 73–90.

9. See Else, "'Imitation' in the Fifth Century."

10. Charles H. Kahn, *The Art and Thought of Heraclitus,* an edition of the fragments with translation and commentary, Fragment XXXII (Cambridge: Cambridge University Press, 1979): 43.

11. See Thomas G. Rosenmeyer, "Gorgias, Aeschylus, and *Apate,*" *American Journal of Philology* 76 (1955): 230. Rosenmeyer writes that the ancients recall that, to read Heraclitus without choking, one had to be a Delian diver.

12. Eric A. Havelock, *Preface to Plato* (Oxford: Basil Blackwell, 1963).

13. Havelock, *Preface,* 3.

14. Plato, *The Republic,* Book III, trans. B. Jowett (Oxford: Clarendon Press, 1888), 79.

15. Rosenmeyer, 226.

16. Plato, *The Republic,* Book X, 316.

17. Havelock, 47. See also Anne Barton, *The Names of Comedy* (Toronto: University of Toronto Press, 1990); and Margaret W. Ferguson, *Trials of Desire: Renaissance Defences of Poetry* (New Haven: Yale University Press, 1983), 8–9.

18. Havelock, 44.

19. Aristotle limits the definition of *mimesis* expressed by Plato, for whom imitation applied to all arts, discourses, and institutions, all figures of some ideal model. For Plato, imitation, aligned with resemblance, is the underlying principle of all things. External reality resembles an ideal something that *is;* a poem, then is an imitation of an imitation. See Ricoeur, *Rule of Metaphor,* 37.

20. "Words may have more than one meaning, but their use in science permits just one.... And it is the division of the sciences that defines this normative usage. Consequently, one and only one literal meaning of *mimesis* is allowed, that which delimits its use in the framework of the *poetical* sciences, as distinct from theoretical and practical sciences" (Ricoeur, *Rule of Metaphor,* 38).

21. Roger Tourangeau, "Metaphor and Cognitive Structure," in *Metaphor: Problems and Perspective,* ed. David Miall (New York: Harvester and Humanities Press, 1982), 33. See also Max Black, "How Metaphors Work," in *On Metaphor,* ed. Sheldon Sacks (Chicago: University of Chicago Press, 1978). Rosemond Tuve tells us that sixteenth- and seventeenth-century writers were quite particular in their notion of delightful metaphor. The reader is expected "to experience greater intellectual pleasure

because simultaneously some true thing is conveyed and a *relatedness* is seen"; *Elizabethan and Metaphysical Imagery* (Chicago: University of Chicago Press, 1947), 122.

22. Ricoeur, "The Metaphorical Process as Cognition, Imagination, and Feeling," in *On Metaphor,* ed. Sacks.

23. Ricoeur, *Rule of Metaphor,* 227. The quotation is from Tzvetan Todorov, *Litterature et Signification* (Paris: Larousse, 1967), 102.

24. de Saussure, *Course in General Linguistics,* ed. Bally, Sechehaye, Riedlinger, trans. Wade Baskin (New York, 1959), 114.

25. Jonathan Culler, *Structuralist Poetics: Structuralism, Linguistics and the Study of Literature* (London: Routledge & Kegan Paul, 1975), 9–10. "Though the rules of *la langue* may be unconscious they have empirical correlates: in the case of language they are manifested in the speakers ability to understand utterances, to recognize grammatically well-formed or deviant sentences, to detect ambiguity, to perceive meaningful relations among sentences, etc. The linguist attempts to construct a system of rules that would account for this knowledge by formally reproducing it. . . . He needs to know, in addition, what (these utterances) mean to speakers of the language, whether they are well formed, whether they are ambiguous and if so in what ways, what changes would alter their meaning or render them ungrammatical. The competence that the linguist investigates is not behaviour itself so much as knowledge which bears upon that behaviour."

26. See Ricoeur, *Rule of Metaphor,* Study 3.

27. Eagleton, *Literary Theory,* 117.

28. Among twentieth-century theorists on this topic, Ted Cohen and Donald Davidson argue that "metaphors mean what the words, in their most literal interpretation mean, and nothing more." Ted Cohen, "Metaphor and the Cultivation of Intimacy," and Donald Davidson, "What Metaphors Mean," both in *On Metaphor,* ed. Sacks.

29. Thomas Sprat, *The History of the Royal-Society of London, For the Improving of Natural Knowledge* (London, 1667); repr. in *English Science, Bacon to Newton,* ed. Brian Vickers (Cambridge: Cambridge University Press, 1987), 171.

30. John Locke is even more decided on this issue: "we must allow that all the art of rhetoric, besides order and clearness, all the artificial and figurative application of words eloquence hath invented, are for nothing else but to insinuate wrong ideas, move the passions, and thereby mislead the judgment, and so indeed are perfect cheats," *Essay On Human Understanding* (1689), Book 3, chapter 10. *The Works of John Locke* 2 (London, 1823; repr. Germany: Scientia Verlag Aalen, 1963), 288.

31. Richard Waswo, *Language and Meaning in the Renaissance* (Princeton: Princeton University Press, 1987). In describing the paradigmatic shift from a relational to a referential semantics, Waswo writes,

> to call a semantic unit a "sign" is to adopt the "picture" of semantic operations that has lain in Western languages since Plato. A sign is a mark, token, or image of something else; its function is extrinsic, standing for or pointing toward something else; and its existence is posterior to that other thing, a copy or trace or symbol of it. If words are regarded as having meaning as "signs," the picture of the world thereby assumed is that painted by ontological and epistemological dualism, in which language becomes merely an imitation or copy of "reality." This picture has been sufficiently falsified in a great many domains of contemporary thought . . . to make it genuinely paradoxical that it should continue to prevail in areas of direct concern with the meaning of language. (4)

For a more comprehensive understanding of imitation theory in representation, see E. H. Gombrich, *Art and Illusion,* 395.

32. Waswo, *Language and Meaning*, 6.
33. Ibid., 33.
34. See Thomas M. Greene, *The Light in Troy*. (New Haven: Yale University Press, 1982), and David Quint, *Origin and Originality in Renaissance Literature: Versions of the Source* (New Haven: Yale University Press, 1983). Quint notes the many variations in thought among the humanist philologists concerning this issue. Pico della Mirandola cautions *against* historical relativism, and the inseparability of style and meaning, asserting philosophy's claim over rhetoric to reveal transcendent truths. See also Quirinus Breen, "Giovanni Pico Della Mirandola on the Conflict of Philosophy and Rhetoric," *Journal of the History of Ideas* (1952): 384–412.
35. Waso, 140.

## 2. *COPIA VERBORUM:* THE MAKER'S KNOWLEDGE IN RENAISSANCE POETICS AND RHETORIC

1. Charles B. Schmitt, *The Cambridge History of Renaissance Philosophy* (Cambridge: Cambridge University Press, 1988), 4.
2. Ibid., 177.
3. Ibid., 137. Schmitt further writes, "we might even wonder whether this may provide a lesson for the present and for the future. If philosophers were to pay greater attention to the humanities, as a few of them have done in our century, this might be beneficial not only for the humanities and humanist scholarship but also for philosophy and for a more complete and more balanced understanding of our world and experience" (137).
4. Pamela O. Long, "Humanism and Science," in *Renaissance Humanism: Foundations, Forms, and Legacy*, vol. 3: *Humanism and the Disciplines*, ed. Albert Rabil, Jr. (Philadelphia: University of Pennsylvania Press, 1988). See also Hans Blumenberg, *The Genesis of the Copernican World*, trans. Robert M. Wallace (Cambridge, Mass.: MIT Press, 1987), and Denise Albanese, *New Science, New World* (Durham, N.C.: Duke University Press, 1987).
5. Paolo Rossis, *Philosophy, Technology, and the Arts in the Early Modern Era*, trans. Salvator Attanasio, ed. Benjamin Nelson (New York: Harper & Row, 1970), 175.
6. Ibid., 176.
7. Long, 487.
8. George A. Kennedy, *Classical Rhetoric and Its Christian and Secular Tradition* (Chapel Hill: University of North Carolina Press, 1980). See also L. D. Reynolds and N. G. Wilson, *Scribes and Scholars: A Guide to the Transmission of Greek and Latin Literature* (Oxford: Clarendon Press, 1968), 101–7. The quotation from Sidney is taken from his *Defence of Poetry*, ed. J. A. Van Dorsten (Oxford: Oxford University Press, 1966), 23.
9. William and Martha Kneale, *The Development of Logic* (Oxford: Clarendon Press, 1962), 229. "When Aristotle's non-logical works were translated, they were at first suspect because they came into circulation with the monopsychist interpretation of the Arabic philosopher Averroes, who lived and worked in Spain. But in the course of the thirteenth century they were freed from suspicion and reconciled with Christianity by Albert the Great and St. Thomas Aquinas."
10. Ibid., 24.
11. Aristotle, *Prior and Posterior Analytics*, ed. John Warrington (London: J. M. Dent & Sons, 1964), 3.
12. Quoted from Helen Vasiliou Martin-Trigona, "Logical Proof and Imaginative

Reason in Selected Speeches of Francis Bacon," (Ph.D. diss., University of Illinois, 1967), 45–46.

13. J. P. Mullally, *The Summulae Logicales of Peter of Spain* (Notre Dame: University of Notre Dame Press, 1945), lxxviii.

14. Martin Elsky, *Authorizing Words: Speech, Writing, and Print in the English Renaissance* (Ithaca: Cornell University Press, 1989), 20.

15. Thomas Aquinas, *Summa Theologica,* trans. Fathers of the English Dominican Province (New York: Benziger, 1947), 1:90.

16. Thomas Aquinas, *Commentary on Aristotle's "On Interpretation,"* (*Peri Hermeneias,* comm. Aquinas and Cajetan), trans. Jean T. Oesterle (Milwaukee: Marquette University Press, 1962), 24.

17. Ibid., 62.

18. Walter Ong, *Ramus, Method, and the Decay of Dialogue,* (Cambridge, MA: Harvard University Press, 1958), 55. "Great as it was, his posthumous reputation as a logician was second to his reputation as a physician—a practitioner both of medicine and of all sorts of physical experimentation, including black magic."

19. Peter of Spain, *Summulae Logicales,* ed. I. M. Bochenski (Turin, 1947), trans. VI, 6.01, p. 57.

20. *Summulae Logicale,* trans. VI, 6.03, pp. 57–58.

21. Ong, 71.

22. Elsky, 23.

23. Ong, *Ramus,* 73–74. "Aquinas takes a rather dim view of discursive reason or ratiocination as compared to sheer understanding, which in its pure state would be intuitive."

24. Ong notes that this emphasis on reason, the ratiocinative process, leads into the eighteenth-century world of reason. "Logic is a study of the reflection of this material world—the world with which man is directly confronted—in the structures of the mind" (74). This Ramistic idea will have interesting repercussions in the work of Milton (see chapter 6).

25. See Charles Fantazzi, *Juan Luis Vives' in Pseudodialecticos: A Critical Edition* (Dordrecht, 1979).

26. Erasmus, *De ratione studii ac legendi interpretandique auctores,* in *Collected Works of Erasmus,* vol. 24, *Literary and Educational Writings* 2, ed. Craig R. Thompson (Toronto: University of Toronto Press, 1978), 666.

27. Terence Cave, *The Cornucopian Text* (Oxford: Clarendon Press, 1979), 21.

28. Erasmus, *De ratione studii,* 669.

29. Cave, 111.

30. Jerry H. Bentley, *Humanists and Holy Writ* (Princeton: Princeton University Press, 1983), 63.

31. For a more complete articulation of this idea, see Cave, 86.

32. On Erasmus' revision of scholastic methodology, see Marjorie O'Rourke Boyle, *Erasmus on Language and Method in Theology* (Toronto: University of Toronto Press, 1977), 124.

33. See Waswo, 223.

34. The Bodleian Library holds the editions of 1513, with three additional editions acquired over the next six years (1514, 1517, 1519), four editions in the 1520s (1521, 1522, 1526, 1528), and four editions in the 1530s (1532, 1535, 1538, 1540). In 1556, the first edition was published in London. Except for two additional editions (1569, 1573) the work was not issued until, interesetingly enough, 1650, with four editions published in the 1660s (1660, 1662, 1664, 1668). Erasmus's *De ratione studii* was issued nine times from 1512 to 1526 (1512, 1513, 1514, 1518, 1519, 1521, 1522, 1524, 1526).

35. Wilson, *The Rule of Reason, conteinying the arte of Logique, set forth in Englishe* (London, 1551), and *The arte of rhetorique, for the use of all such as are studious of eloquence* (London, 1553), both printed by Richard Grafton. All quotations are taken from these two editions.

36. Howell, *Logic and Rhetoric in England, 1500–1700* (Princeton: Princeton University Press, 1956), 23.

37. Sidney, *A Defence of Poetry* (1595). All references are taken from this work and cited as *Defence* in parenthesis in the text.

38. See Ferguson, 15, who argues that Renaissance authors acknowledge their status as members of the body politic, what Sidney calls a "predicament of relation" (17).

39. See O. B. Hardison, "The Two Voices of Sidney's *Apology for Poetry*," *ELR* 2 (1972), who argues that Sidney uses his second voice, his philosopher's voice, to undermine the claims made earlier by his poet's voice. I would assert that the two voices are not antithetical, in the same way that Bacon can claim that his scientific method combines truth with usefulness, theory with practice.

40. Ferguson, 155.

41. Bacon, *The Essays,* ed. John Pitcher (Harmondsworth, Middlesex: Penguin Books, 1985), 61. See also, Kenneth Alan Hovey, "'*Mountaigny* Saith Prettily': Bacon's French and the Essay," in *PMLA* 106 (1991): 71–82. Hovey argues that both Bacon and Montaigne address a threefold division of philosophers, the dogmatics, the Academics, and the skeptics. "Both Montaigne and Bacon . . . regard the essay as a form conveyed to them most directly by Seneca and Plutarch but originally designed by the pre-Socratics to express a peculiarly nonpositive, nonnegative philosophy, a way of thought both Renaissance essayists support as part of a supposed third school," 74.

## 3. Francis Bacon: The "Restauration"

1. Bacon, *The Philosophical Works of Francis Bacon,* ed. John M. Robertson (New York, 1905). I quote from Robertson's edition of the *Novum Organon* (cited in the text as *NO*), *De Dignitate et Augmentis Scientarum* (*DAS*) and *De Sapientia Veterum* (*DSV*). I use Arthur Johnston's 1974 Oxford edition of *The Advancement of Learning* (*ADV*) and the *New Atlantis* (*NA*). The quotation concerning truth is from Bacon's essay, "Of Truth," in *The Essays,* ed. John Pitcher (Harmondsworth, Middlesex, 1985), 61.

2. Benjamin Farrington, *The Philosophy of Francis Bacon* (Liverpool: Liverpool University Press, 1970), 13.

3. Farrington, 13–15. See also Martin Elsky, *Authorizing Words: Speech, Writing, and Print in the English Renaissance* (Ithaca: Cornell University Press, 1989). Elsky argues similarly that Bacon wishes to persuade his uncle to appoint him head of some institution of learning in order to implement his reform of science and philosophy. "An administrative career would have been more socially acceptable to a member of the governing class who saw his proper place in political activity" (191).

4. See Bacon's correspondence during this time, in *The Letters and Life of Francis Bacon,* published as volumes 8 to 14 in *The Works of Francis Bacon,* 14 vols., ed. James Spedding, R. L. Ellis, and D. D. Heath (reprint, New York: Garrett Press, 1968). Bacon's correspondence in the years 1588–1603 shows that he vacillates constantly in his desire for either an active public life or the more contemplative life of a scholar. Although he threatens his uncle that he will become "a sorry book-maker" if a Crown position is not attained, he writes to Essex of his plans to retire to Cambridge and spend his life in studies and contemplations, without looking back. (8:291).

5. Farrington, 16.

6. For the dating of these texts, see Farrington, 11. I use Farrington's translation (cited in the text as *MBT*), *Thoughts and Conclusions* (*TC*), and *The Reputation of Philosophies* (*RP*). The first two works were not published until 1653.

7. Bacon, *The Great Instauration,* author's preface, in *The Philosophical Works of Francis Bacon,* ed. John M. Robertson (New York, 1905), 244.

8. Rossi, *Francis Bacon: From Magic to Science* (Chicago: University of Chicago Press, 1968), 93.

9. Briggs, *Francis Bacon and the Rhetoric of Nature* (Cambridge, Mass.: Harvard University Press, 1989), 32.

10. J. Spedding, ed. *The Letters and Life of Francis Bacon* (London, 1890), 4:141.

11. Five of Bacon's thirty-nine philosophical works are concerned with the hidden wisdom of fables: *Cogitationes De Scientia Humana* (written in 1604), *The Advancement of Learning* (1605), *De Sapientia Veterum* (1609), *De Dignitate et Augmentis Scientiarum* (1623), and *De Principiis Atque Originibus* (1623).

12. Bacon, *De Dignitate et Augmentis Scientiarum,* in *The Philosophical Works of Francis Bacon,* ed. John M. Robertson (New York, 1905). I use Arthur Johnston's edition of *The Advancement of Learning* (Oxford: Clarendon Press, 1974). Quotations taken from the English translation of *De Augmentis* will be indicated as *DAS* in the text, and from the English *Advancement of Learning* as *ADV.*

13. See Charles Whitney, *Francis Bacon and Modernity* (New Haven, 1986). The sons of science are "to deliver knowledge 'in the same method in which it was invented,' a method in which 'ornaments of speech, similitudes, treasury of eloquence and such like emptiness' are 'utterly dismissed' . . . Bacon's induction by negation restricts the active, similitude-generating faculties of the mind, according to which men wrongfully 'divine of the new with an imagination preoccupied and coloured by the old'" (145). Lisa Jardine, *Francis Bacon: Discovery and the Art of Discourse* (Cambridge: Cambridge University Press, 1974), argues that, for Bacon, only the inductive method has any privileged status. "For him all teaching was insinuative; all presentation is misrepresentation to some specified end" (75).

14. Julian Martin, *Francis Bacon, the State, and the Reform of Natural Philosophy* (Cambridge: Cambridge University Press, 1992), 173. Martin also argues that Bacon builds a paternal governance in his province of knowledge by harnessing "Tudor strategies of bureaucratic state management, late-Elizabethan political anxieties and social prejudices, and a principal intellectual resource of the landed gentry—the science of the common law" (173). My argument adds that Bacon used pre-Socratic rhetorical theory to buttress his ideas that scientific knowledge must be protected from the uninitiated.

15. Morris W. Croll, "Attic Prose: Lipsius, Montaigne, Bacon," in *Style, Rhetoric, and Rhythm: Essays by Morris W. Croll,* ed. J. Max Patrick and Robert O. Evans (Princeton: Princeton University Press, 1966), 163. Originally published in *Schelling Anniversary Papers by His Former Students,* New York: The Century Co., 1923, 117–50. According to the editors, Croll defines Anti-Ciceronian, or "Attic," style as a reaction against the artificialities of a sixteenth-century emphasis on words and forms, "in each, content and the mind in the process of thinking took precedence over conventional form and the rigidities of genre and tradition" (163).

16. Croll, "The Baroque Style in Prose," in *Style, Rhetoric, and Rhythm: Essays by Morris W. Croll,* 222, first published in *Studies in English Philosophy: A Miscellany in Honor of Frederick Klaeber,* ed. Kemp Malone and Martin B. Ruud (Minneapolis: University of Minnesota Press, 1929).

17. Arthur Johnston, ed., *The Advancement of Learning and New Atlantis* (Oxford,

1974), 281. Johnston identifies this secret, acroamatic method with the alchemists and magicians Agrippa and Cardan. I argue that Bacon is referring to his own reformed, insinuative method, which uses similitudes and enigmas to strengthen the intellectual capacity of the receiver, thereby furthering the progression of knowledge.

18. In contrast, knowledge delivered in methods is illustrated with examples and digested into a methodical system. Of course, Bacon never fully succeeds in producing such an aphoristic induction. In his later work, he becomes more "magistral" in his mode of delivery. Whereas in the earlier *Essays* of 1597, Bacon is content to list a series of maxims with few or no connective passages, in the revised *Essays* of 1612 and 1625, he has come to realize that the author cannot rely completely on the reader's imagination; some connective and interpretive tissues are necessary.

19. See Terence Cave, *The Cornucopian Text*, 86.

20. Farrington calls this the "intellectualising of the industrial process, the emergence of science proper out of a merely economic activity" (119).

21. Ricoeur, *Rule of Metaphor*, 6. Ricoeur is, of course, referring to Aristotle's definition of *mimesis* in poetry. My intent, however, is to show how this creative spirit in Aristotelian *mimesis* finds its way into seventeenth-century scientific writing.

22. See Ricoeur, *Rule of Metaphor*, 39. While Sir Philip Sidney distinguishes between *eikastike* and *phantastike* poetry, ultimately he decides that the poet is limited only by "the zodiac of his own wit." Sidney resolves the moral issues of representation, and the lying counterfeit, by centering, pragmatically, upon the intention of the imitator. *A Defence of Poetry*, ed. Jan Van Dorsten, 24, 25, 54. On the Renaissance understanding of *phantasia* in bringing language into existence, see Eugene Vance, *Mervelous Signals: Poetics and Sign Theory in the Middle Ages* (London, 1986), 333. See also Margaret W. Ferguson, *Trials of Desire: Renaissance Defenses of Poetry*.

23. Quintilian, *Institutio Oratoria*, VII, x, 9, trans. H. E. Butler (Cambridge, Mass.: Loeb Classical Library, 1958). "What painter has ever been taught to reproduce everything in nature? But once he has acquired the general principles of imitation, he will be able to copy what ever is given him. What vase-maker is there who has not succeeded in producing a vase of a type which he had not previously seen?"

24. Erasmus, *De ratione studii*, in *Collected Works of Erasmus*, ed. Craig R. Thompson, vol. 24, 666.

25. Augustine, *Confessions*, trans. R. S. Pine-Coffin (London: Penguin Books, 1961), VII, xvii, 151.

26. Augustine, *De Magistro*, trans. George G. Leckie (London: Appleton-Century Company, 1938), 5.

27. Augustine, *De Doctrina Christiana*, trans. D. W. Robertson, Jr. (New York: Macmillan, 1958), II, i, 34.

28. Morris W. Croll, "'Attick Prose' in the Seventeenth Century," in *Style, Rhetoric, and Rhythm: Essays by Morris W. Croll,* originally published in *Studies in Philology* 18 (April 1921): 79–128. Croll writes that Isocratean rhetoric is more suited to the public assemblage and "Attic prose" to the "inward ear of the solitary reader" (56).

## 4. Bacon's Politics of Allegory

1. This frustration is echoed in Bacon; his division of memory into prenotional and emblematic functions partially repairs this scission of mental images and verbal representation.

2. Elsky, 170.

3. See Sidney Warhaft, "The Providential Order in Bacon's New Philosophy,"

*Studies in the Literary Imagination* 4 (1971); and Margreta de Grazia, "The Secularization of Language in the Seventeenth Century," *Journal of the History of Ideas* 41 (1980). De Grazia's argument starts from the premise that the seventeenth century, in general, is characterized by its distrust of ordinary language, due to the growing recognition that language is not the gift of an omniscient and omnipresent God, but the suspect tool invented by the feeble intellect of man. "Without any prerogative to reproduce or translate God's Word as imprinted in nature, in the Scriptures, and in the heart, words were of limited use and value." Because of this, a philosophical grammar or language became a primary concern with Wilkins, Ward, and other language theorists of the Royal Society. G. A. Padley confirms this view, saying that by the time of Hobbes and Locke, the separation between word and thing reaches a dangerous extreme; *Grammatical Theory in Western Europe, 1500–1700* (Cambridge: University of Cambridge Press, 1985), 141–42. Padley, however, cautions the reader that, because of his distrust of words, and his suspicion of a reliable correspondence between sign and referent, Bacon did not envision a philosophical language at all, especially in the mode of Wilkins. On this point, see also Vivian Salmon, "Language-Planning in Seventeenth-Century England, Its Context and Aims," in *The Study of Language in Seventeenth-Century England, Amsterdam Studies in the Theory and History of Linguistic Science,* ed. E. F. K. Koerner, Series 3, xvii (Amsterdam, 1979): 129–56.

4. See Pagel, *Paracelsus: An Introduction to Philosophical Medicine in the Era of the Renaissance* (Basel and New York: S. Karger, 1958). Pagel marvels at the odd combination found in Paracelsus of the magical, the fantastical, and the scientifically progressive.

5. William Sessions, "Bacon and Herbert and an Image of Chalk," in *'Too Rich to Clothe the Sunne': Essays on George Herbert* (Pittsburgh: University of Pittsburgh Press, 1980).

6. Elsky, 170.

7. Ibid., 177.

8. Judith H. Anderson, *Words That Matter: Linguistic Perception in Renaissance English* (Stanford: Stanford University Press, 1996), 17.

9. Reiss, *The Discourse of Modernism* (Ithaca: Cornell University Press, 1982), 216.

10. Annabel Patterson, *Fables of Power: Aesopian Writing and Political History* (Durham, NC: Duke University Press, 1991), 55.

11. Jonathan Goldberg, *Voice Terminal Echo, Postmodernism and English Renaissance Texts* (New York and London: Methuen, 1986).

12. Quilligan, *The Language of Allegory* (Ithaca: Cornell University Press, 1979), 29.

13. de Man, "The Rhetoric of Temporality," in *Blindness and Insight: Essays in the Rhetoric of Contemporary Criticism,* ed. Wlad Godzich (London: Methuen, 1983).

14. Ibid., 188.

15. Ibid., 207.

16. In an important article on the subject, Stephen H. Daniel articulates how the philosophic use and study of myth is a basic component of Bacon's new instauration of learning. In his new theory of discovery or logic of problem-solving, Bacon relies on the literary character of both nature and our experience of it: "Myth and metaphoric writings are especially appropriate for the description of the world as created and yet still resistent to closure of meaning; i.e., to describe the world in mythic terms reveals how the referential meaning or representational claims of any account of the world must remain indeterminate: closure of meaning is possible only within the domain of the signifying-signified." See his "Myth and the Grammar of Discovery in Francis Bacon," *Philosophy and Rhetoric* 15 (1982): 219–37. My own understanding of Bacon's *literata experientia* as a necessary method for the discovery of knowledge preceding the *novum organon* or interpretation of nature is similar to that of Daniel.

17. Ibid., 224.
18. Ibid., 219. This kind of epistemology relates to the third school of philosophers (the pre-Socratics) praised by Bacon in his "Praefatio" to *The Great Instauration*. These most ancient of the Greeks are held to be superior to the dogmatics (the Academics) and the skeptics. On this threefold division of philosophy, see Kenneth Alan Hovey, "'*Mountaigny* Saith Prettily': Bacon's French and the Essay" in *PMLA* 106 (1991): 73.
19. Bacon, *Valerius Terminus*, in *The Philosophical Works of Francis Bacon*, ed. Robertson, 192.
20. Jon Whitman, *Allegory: The Dynamics of an Ancient and Medieval Technique* (Cambridge, Mass.: Harvard University Press, 1987).
21. Ibid., 263.
22. James L. Calderwood, *Metadrama in Shakespeare's 'Henriad': Richard II to Henry V* (Berkeley: University of California Press, 1979).
23. Clark Hulse, "Spenser, Bacon and the Myth of Power" in *The Historical Renaissance: New Essays on Tudor and Stuart Literature and Culture*, ed. Heather Dubrow and Richard Strier (Chicago: University of Chicago Press, 1988). See also David Norbrook, *Poetry and Politics in the English Renaissance* (London: Routledge & Kegan Paul, 1984). Both claim that the Tudor and Stuart courts were increasingly reliant on masques as a form of entertainment and a justification of their own ruling strategies.
24. Hulse, 338.
25. Elsky, 175. "While Herbert's hieroglyphics, for example, frame an inherent tension between divine writing and the human voice that must match it, Bacon's hieroglyphs dissolve that tension and preclude the failure of signification that may arise when the human voice falters. The system of correspondence that can go awry, thereby issuing words that cannot match their referents, is exactly what Bacon is trying to discard." For another view, see James Stephens, *Francis Bacon and the Style of Science* (Chicago: University of Chicago Press, 1975). Stephens argues that Bacon's preoccupation with myth is in keeping with his basic psychology of discovery. A myth or parable, by its nature, presents a problem for the scientific interpreter, a problem which needs to be solved.
26. Daniel, 233. Daniel further states that "by means of drawing a curtain, Bacon intended not to obscure truths already known (and to be protected from vulgar eyes) but rather to provide the cautionary foundations from which to suggest further philosophic explorations" (232). A less sympathetic reading of Bacon's use of fable is provided by Clark Hulse: "Bacon sets up a three-fold division of human time: an antique age of oblivion, an age of fable, and a modern age of records. Yet, rather than define fable more closely, he collapses the middle stage, making it a mere 'vaile' between things known and things lost, a veil that has not been placed there deliberately, but has 'interposed itselfe'. In this way, the deliberate act of meaning by the poetic transmitter of myth is effaced, and the veil is defined as a condition intrinsic both to philosophical parable itself and to our historical distance from it" (335).
27. Jardine, 193.
28. Bacon uses the myth of the birth of Cupid to endorse two opposing views on the accessibility of natural knowledge. In *De Sapientia Veterum*, the birth symbolizes the inaccessibility of the laws of nature. In the later *De Principiis Atque Originibus*, Bacon uses the same myth to argue the opposite viewpoint.
29. This conforms to Donald Davidson's rule of metaphor in his article "What Metaphors Mean," in *On Metaphor*, ed. Sacks, 33. "The key words in a metaphor have two different kinds of meaning at once, a literal and a figurative meaning . . . there must be a rule which connects the two meanings. . . . The rule says that in its

metaphorical role the word applies to everything that it applies to in its literal role, and then some."

30. Ted Cohen, "Metaphor and the Cultivation of Intimacy," in *On Metaphor,* ed. Sacks (Chicago: University of Chicago Press, 1978), 4.

31. See Ricoeur, "The Metaphorical Process as Cognition, Imagination, and Feeling," in *On Metaphor.* "Poetic language is no less about reality than any other use of language but refers to it by the means of a complex strategy which implies a suspension and seemingly an abolition of the ordinary reference attached to descriptive language," 153.

32. "The notion of poetic image and of poetic feeling has to be construed in accordance with the cognitive component understood itself as a tension between congruence and incongruence at the level of sense, between epoche and commitment at the level of reference"; Ricoeur, "Metaphorical Process," 158.

## 5. Wise Men's Counters: Visual and Verbal Knowledge in Hobbes and Boyle

1. Antonio Pérez-Ramos, "Bacon's Legacy," in *The Cambridge Companion to Bacon,* ed. Markku Peltonen (Cambridge: Cambridge University Press, 1996), 312.

2. Robert Foster Jones, in his articles "Science and English Prose Style in the Third Quarter of the Seventeenth Century," *Publication of the Modern Language Association* 45 (1930), and "Science and Language in England of the Mid-Seventeenth Century," *Journal of English and Germanic Philology* 31 (1932), was the first to argue persuasively the extent of science's influence on the style of English prose during the Restoration, mainly through the writings of Boyle, Sprat, and Glanville following the example set by Bacon. See also "The Background of *The Battle of the Books,*" *Washington University Studies* 7 Humanistic Series II (1920), on the debate between the ancient and moderns and "Science and Criticism in the Neo-Classical Age of English Literature," *Journal of the History of Ideas* 1 (1940), on Bacon's influence on literary criticism of the Restoration, especially that of Dryden.

3. One group of scientists met in London in 1645, later in Gresham College, and included John Wallis, the mathematician, John Wilkins, and Jonathan Goddard. Others, such as Boyle and Petty, were connected with the Puritan educational reformer Samuel Hartlib. In fact, in May 1654, Hartlib wrote to Boyle about a visit he had made to Lambeth Marsh, "to see part of that foundation or building, which is designed for the execution of my lord *Verulams' New Atlantis,*"; Boyle, *Works* 6:86, 88.

4. I am responding to earlier work on the subject by Lisa Jardine in her book, *Francis Bacon: Discovery and the Art of Discourse* (Cambridge: Cambridge University Press, 1974); Charles Whitney, in *Francis Bacon and Modernity* (New Haven: Yale University Press, 1986); and Stephen Daniel, "Myth and the Grammar of Discovery in Francis Bacon," *Philosophy and Rhetoric* 15 (1982), to name just a few. Research in this field includes the work of Robert Markley, "Robert Boyle on Language," *Studies in Eighteenth-Century Culture* 12 (1985); and "Objectivity as Ideology: Boyle, Newton and the Language of Science," *Genre* 16 (1983); James Paradis, "Montaigne, Boyle, and the Essay of Experience," *One Culture: Essays in Science and Literature,* ed. George Levine (Madison: University of Wisconsin Press, 1987); William Youngren, "Generality, Science and Poetic Language in the Restoration," *English Literary History* 35 (1968); and Roger Pooley, "Language and Loyalty: Plain Style at the Restoration," *Literature and History* 6 (1980). Vivian Salmon's work on Francis Lodwick and the study of language in the seventeenth century is also invaluable (*The Works of Francis*

*Lodwick: A Study of His Writings in the Intellectual Context of the Seventeenth Century* (London: Longman, 1972), and *The Study of Language in Seventeenth-Century England* (Amsterdam, 1979). I would like to contribute to this body of research by looking at the scientific writings, and religious polemic, of various late seventeenth-century members of the Royal Society. These writers include Thomas Sprat, Robert Boyle, Thomas Glanville, Joseph Webster, and Seth Ward.

5. Brian Vickers, ed., *English Science, Bacon to Newton* (Cambridge: Cambridge University Press, 1987), 227n20.

6. Sprat, *History of The Royal Society* (1667), in Vickers, ed., *English Science, Bacon to Newton*, 160.

7. "Now men are generally weary of the *Relicks of Antiquity,* and satiated with *Religious disputes;* now not only the *eyes* of men but their *hands* are open, and prepar'd to *labour.* Now there is a universal *desire* and *appetite* after *knowledge,* after the peaceable, the fruitful, the nourishing *Knowledge,* and not after that of antient Sects, which only yielded hard indigestible arguments, or sharp *contentions* instead of *food;* which when the minds of men requir'd *bread* gave them only a *stone,* and for *fish* a *serpent*"; Sprat, 173.

8. I am using the edition by W. G. Pogson Smith, of Hobbes' *Leviathan,* reprinted from the edition of 1651 (Oxford: At the Clarendon Press, 1909). References are taken from this text and unless otherwise specified are cited as *Leviathan* in parentheses after the quotation.

9. See Steven Shapin and Simon Schaffer's *Leviathan and the Air-Pump: Hobbes, Boyle, and the Experimental Life* (Princeton: Princeton University Press, 1985), 7.

10. See Quilligan, 1979.

11. G. A. Padley argues that by the time of Hobbes, the separation between word and thing had reached a dangerous extreme, 141–42.

12. David Johnston disputes the claim that Hobbes was a forerunner of scientific positivism and liberalism. By focusing on what he calls the rhetoric of *Leviathan,* Johnston restores an awareness of "the roots of Hobbes' thinking in the humanist and polemical traditions of the Renaissance . . . showing how these elements shaped his argumentation in *Leviathan*"; *The Rhetoric of Leviathan: Thomas Hobbes and the Politics of Cultural Transformation* (Princeton: Princeton University Press, 1986), ix. Johnston's approach is similar to mine regarding Bacon, namely to demonstrate that Hobbes' method owes as much to sixteenth-century modes of thought as it does to the seventeenth-century scientific outlook.

13. See Vivian Salmon's *The Works of Francis Lodwick* (London: Longman, 1972) and *The Study of Language in Seventeenth-Century England* (Amsterdam, 1979).

14. John Wilkins, "Dedicatory Epistle," in *An Essay Towards a Real Character and a Philosophical Language* (London, printed for S. Gellibrand, and for John Martyn, Printer to the Royal Society, 1668; A Scolar Press Facsimile, 1968), i.

15. See Padley, 329; and Paolo Rossi, *Francis Bacon: From Magic to Science* (Chicago: University of Chicago Press, 1968), 145.

16. Padley claims that "more recently it has been objected that not only did Bacon not have in mind a 'philosophical language' of the type later produced by Wilkins and others, but merely a 'universal character' or alphabet of symbols, but also his distrust of words did not allow him to envisage the possibility of a reliable correspondence between word and referent," 330. Charles Webster, in his *The Great Instauration: Science, Medicine and Reform, 1626–1660* (London: Holmes & Meier, 1975), also warns that there is little uniformity of ethos among the members of the Royal Society, each member citing Bacon as a source without completely understanding his attitude toward language and science. Salmon also states that "Bacon

still did not demand in completely unequivocal terms the creation of a new language for scientific communication," 146.

17. Robert Boyle, *The Christian Virtuoso* (London, by T.H.R.B., Fellow of the Royal Society, 1690), sig. A5v.

18. Boyle, *Certain Physiological Essays, The Works of the Right Honourable Robert Boyle,* ed. Thomas Birch, 6 vols. (London, 1772). Individual works will be cited by the titles along with the volume and page number of this edition. John T. Harwood's "Science Writing and Writing Science: Boyle and Rhetorical Theory," in *Boyle Reconsidered*, ed. Michael Hunter (Cambridge: Cambridge University Press, 1994) is a useful study of the effect of rhetorical theory on Boyle's epistemology, as is Peter Dear's "Totius in Verba: Rhetoric and Authority in the Early Royal Society," *Isis* 76 (1985): 145–61. See also Jan W. Wojcik's *Robert Boyle and the Limits of Reason* (Cambridge: Cambridge University Press, 1997) for a full discussion of Boyle's defense of the style of Scripture as a beneficial tool for the development of reason, despite its obscurity, and unmethodical, self-contradictory mode of logic.

19. According to J. Paul Hunter ("Robert Boyle and the Epistemology of the Novel," *Eighteenth-Century Fiction* 2 [July 1990]: 275–82), Boyle was a sensitive, visceral reader. He was opposed to the reading of romances for a deeply personal reason: they mesmerized his imagination. See also Hunter's important study, *Before Novels: The Cultural Contexts of Eighteenth-Century English Fiction* (New York, 1990), in which he argues that Boyle played an important role in creating an audience for fiction, an irony, given Boyle's seeming distrust of fictional forms. My argument, in contrast, states that Boyle had a great respect for fictional forms in manipulating reader response, hence his constant quest for the most effective discursive form in the transmission of what he considered proven scientific truths.

20. John Beale encouraged Boyle to model his career after Erasmus, who published in short works and who preserved his letters for publications. See Michael Hunter, *Science and Society in Restoration England* (Cambridge: Cambridge University Press, 1981), 194–97; and *Establishing the New Science: The Experience of the Early Royal Society* (Woodbridge, 1989).

21. Sargent, "Scientific Experiment and Legal Expertise: The Way of Experience in Seventeenth-Century England," *Studies in History and Philosophy of Science* 20:1 (March 1989): 28. See Sargent's *The Diffident Naturalist: Robert Boyle and the Philosophy of Experiment* (Chicago: University of Chicago Press, 1995); Sargent's "Learning from Experience: Boyle's Construction of an Experimental Philosophy, in *Boyle Reconsidered*, ed. Michael Hunter (Cambridge: Cambridge University Press, 1994); and "Robert Boyle's Baconian Inheritance: A Response to Laudan's Cartesian Thesis," *Studies in the History and Philosophy of Science* 17 (1986): 469–86. Also useful is Daniel R. Coquillette's *Francis Bacon,* in the series *Jurists: Profiles in Legal Theory* (Edinburgh, 1992); Brian P. Levack, *The Civil Lawyers in England* (Oxford, 1973); Wilfred R. Prest, *The Inns of Court Under Elizabeth I and the Early Stuarts* (Totowa, N.J.: Rowan and Littlefield, 1972); W. C. Richardson, *A History of the Inns of Court* (Baton Rouge, 1975); and John H. Langbein, *Prosecuting Crime in the Renaissance* (Cambridge, Mass., Harvard University Press, 1974).

22. Canny, *The Upstart Earl: A Study of the Social and Mental World of Richard Boyle, First Earl of Cork* (Cambridge: Cambridge University Press, 1982), 147.

23. Webster, *The Great Instauration: Science, Medicine and Reform, 1626–1660* (New York: Holmes & Meier, 1975). See also Malcolm Oster, "Virtue, Providence and Political Neutralism: Boyle and Interrugnum Politics," in *Boyle Reconsidered*, ed. Michael Hunter (Cambridge: Cambridge University Press, 1994).

24. Jacob, *Robert Boyle and the English Revolution: A Study in Social and Intellectual*

*Change* (New York, 1977). See also T. C. Barnard, *Cromwellian Ireland: English Government and Reform in Ireland, 1649–60* (Oxford, 1975); J. R. Jacob and Margaret C. Jacob, "The Anglican Origins of Modern Science: The Metaphysical Foundations of the Whig Constitution," *Isis* 71 (1980): 251–67; Margaret C. Jacob, *The Cultural Meaning of the Scientific Revolution* (Philadephia, 1988); and Canny, *Upstart Earl,* 144.

25. This central element of the thesis has been called into question by John Morgan, "Puritanism and Science: A Reinterpretation," in *Historical Journal* 20 (1979): 535–60. See also Margo Todd's *Christian Humanism and the Puritan Social Order* (Cambridge, 1987), a survey of humanism's impact on the Protestant social ethos.

## 6. The Figure in the Pool: Milton's Epistemology of Nature

1. Slaughter, *Universal Languages and Scientific Taxonomy in the Sixteenth Century* (Cambridge: Cambridge University Press), 11. The quotation from Alberti is taken from *De Pictura,* Book II, section 26, in *On Painting and On Sculpture,* trans. Cecil Grayson (London: Phaidon Press, 1972), 61 62.

2. Ibid., 11.

3. John D. Lyons, "Speaking in Pictures, Speaking of Pictures: Problems of Representation in the Seventeenth Century" in *Mimesis: From Mirror to Method, Augustine to Descartes,* ed. John D. Lyons and Stephen G. Nichols, Jr. (Hanover, NH: University Press of New England, 1982), 167.

4. Gilman, *The Curious Perspective: Literary and Pictorial Wit in the Seventeenth Century* (New Haven: Yale University Press, 1978), 14.

5. Again, I refer the reader to Waswo's comment that in imitation theory, the semantic unit is a mere picture or copy of objective reality, meaning being conferred by reference and semantics proceeding "to debate the nature of the something else referred to. Theory has no need to question the connection, but merely relocates what it is a connection to" (Richard Waswo, *Language and Meaning in the Renaissance,* Princeton: Princeton University Press, 1987, p. 33). My point is that even in this kind of "referential" semantics, the nature of the "signified" object is never decided in such an unproblematic way, as Waswo implies, or divorced from the problem of visual perception. The dilemma posed by visual perception, and the related problem of mental cognition, ensures that the connection between words and things can never go unquestioned, as Waswo asserts.

6. Lyons, "Speaking in Pictures," 169.

7. Antoine Arnauld and P. Nicole, *Logic, or the Art of Thinking,* ed. James Dickoff and Patricia Jones (Indianapolis: Bobbs-Merrill Library of Liberal Arts, 1964), 28.

8. Ibid., 28.

9. Blaise Pascal, *On the Geometrical Mind and the Art of Persuasion* (1657–1658) in *Pascal Selections,* ed. and trans. by Richard H. Popkin (New York: Macmillan, 1989), 180.

10. Ibid., 194.

11. In this kind of study, I limit myself to the attitude displayed by Milton regarding figurative language, imitation and mimicry, artificial resemblances, and mirrored images. I will not be commenting upon the complete linguistic ethos of the poet, or the theological and cosmological systems displayed in his work.

12. Sloan, *Donne, Milton and the End of Humanist Rhetoric* (Berkeley: University of California Press, 1985), 284.

13. In the introduction to his translation of Milton's *The Art of Logic,* Walter Ong

endorses Sloan's view that Milton does not "deviate in any significant way from the logic of Ramus on which it explicitly structures itself" (147). See *A Fuller Course in the Art of Logic* (1672), ed. and trans. by Walter J. Ong and Charles J. Ermatinger, in *Complete Prose Works of John Milton* 8 (New Haven: Yale University Press, 1982): 147.

14. Ong, 202.

15. Ibid., 204.

16. Milton, *The Reason of Church Government Urged Against Prelaty,* in *The Prose Works of John Milton,* vol. II, edited by J. A. St. John (London: George Bell and Sons, 1893), 464.

17. My aim here is to relate the linguistic attitude of Milton to that of Hobbes. Theologically, the two are not in agreement, given the Puritanism of Milton and the religious nominalism of Hobbes. Ultimately, this disagreement demonstrates a fundamental divergence regarding the correspondence of words and things. While Hobbes exhibits a certain confidence in reason's ability arbitrarily to assign words to things, Milton tends to subvert and dismantle the very ratiocinative process which makes such relationships possible. See Stanley Fish, "Reasons that Imply Themselves: Imagery, Argument, and the Reader in Milton's *Reason of Church Government,*" in *Seventeenth-Century Imagery: Essays on Uses of Figurative Language from Donne to Farquhar,* ed. Earl Miner (Berkeley: University of California Press, 1971).

18. *Reason of Church Government,* 464.

19. Ibid., 500.

20. Fish, "Reasons That Imply Themselves," 92–93.

21. Ibid., 101.

22. For a useful discussion of the law of metamorphosis and transferral in mythological language, see Ernest Cassirer, *An Essay on Man* (New Haven: Yale University Press, 1944): "The limits between the different spheres are not insurmountable barriers; they are fluent and fluctuating. There is no specific difference between the various realms of life. Nothing has a definite, invariable, static shape. By a sudden metamorphosis everything may be turned into everything" (81).

23. See Isabel MacCaffrey, *Paradise Lost as 'Myth'* (Cambridge, MA: Harvard University Press, 1959), 208, for an insightful discussion on this theme.

24. John Milton, *Paradise Lost,* in *Paradise Lost and Selected Poetry and Prose,* ed. Northrop Frye (New York, 1951).

25. Milton, *Paradise Regained,* in *Complete Shorter Poems,* ed. John Carey (Essex, 1968).

26. Robert Hoopes, *Right Reason in the English Renaissance* (Cambridge, MA: Harvard University Press, 1962), notes that intuitive right reason is the "intimate impulse whereby God speaks and moves a man to action; a power that fuses thought and action," while discursive reason "depends upon inference from the facts of primary perception, which reason, acting upon these facts, builds into theory and system" (194).

27. Milton, *Areopagitica,* in *Paradise Lost and Selected Poetry and Prose,* ed. Northrop Frye (New York: Holt, Rinehart & Winston, 1951), 498.

28. Milton, *Of Education,* in *Paradise Lost and Selected Poetry and Prose,* ed. Northrop Frye (New York: Holt, Rinehart & Winston, 1951), 439.

29. Ibid., 439.

30. In this sense, Milton's notion of right reason departs from that of the Greeks. Hoopes writes that the "question of whether another authority ought to pass upon reason's ability to confirm the truth it discloses rarely occurred to the Greeks. The truth is 'there' to be discovered, but that which proclaims truth as truth inheres in reason itself."

31. Regina Schwartz, "Through the Optic Glass: Voyeurism and *Paradise Lost,*" in *Desire in the Renaissance: Psychoanalysis and Literature,* ed. Valeria Finucci and Regina

Schwartz (Princeton: Princeton University Press, 1994), 147. See also Schwartz, *Remembering and Repeating: On Milton's Theology and Poetics* (Cambridge: Cambridge University Press, 1988); and "The Toad at Eve's Ear: From Identification to Identity," in *Literary Milton*, ed. Diana Trevino Benet and Michael Lieb (Pittsburgh: Duquesne University Press, 1994). For Freud, see his "Instincts and Their Vicissitudes," in *On Metapsychology: The Theory of Psychoanalysis,* trans. James Strachey, Penguin Freud Library, vol. 2 (London: Penguin Books, 1984).

32. William Shakespeare, *The Riverside Shakespeare* (Boston: Houghton Mifflin Company, 1974).

33. Claudio Guillen, "On the Concept and Metaphor of Perspective," in *Comparatists at Work,* ed. Stephen G. Nichols, Jr. and Richard B. Vowles (Waltham, MA: Blaisdell Publishing Company, 1968), 49. See also Fred Leeman's *Hidden Images: Games of Perception, Anamorphic Art, Illusion,* trans. Ellyn Childs Allison and Margaret L. Kaplan (New York: Harry N. Abrams, 1976); Samuel Y. Edgerton, Jr., *The Renaissance Rediscovery of Linear Perspective* (New York: Basic Books, 1975). On optics, see David C. Lindberg, *John Pecham and the Science of Optics* (Madison: University of Wisconsin Press, 1970); and A. I. Sabra, *Theories of Light from Descartes to Newton* (New York: American Elsevier, 1967).

34. Lyons, "Speaking in Pictures, Speaking of Pictures: Problems of Representation in the Seventeenth Century" in *Mimesis: From Mirror to Method, Augustine to Descartes,* ed. John D. Lyons and Stephen G. Nichols, Jr. (Hanover, NH: University Press of New England, 1982), 167.

35. M. M. Slaughter, *Universal Languages and Scientific Taxonomy in the Sixteenth Century* (Cambridge: Cambridge University Press, 1982), 11.

36. Guillen, 47.

37. Gilman, *The Curious Perspective: Literary and Pictorial Wit in the Seventeenth Century* (New Haven: Yale University Press, 1978), 14. On the relationship between pictorial space and the arts of eloquence, see Michael Baxandall, *Giotto and the Orators* (Oxford: At the Clarendon Press, 1971). Baxandall notes that it is Petrarch who argues that in imitating a literary model, many things should be dissimilar rather than similar. The illusion of verisimilitude should be avoided, for deception and vanity is the result (33). On the seductiveness of images, see David Freedberg's *The Power of Images: Studies in the History and Theory of Response* (Chicago: University of Chicago Press, 1989).

38. Baltrusaitis, *Anamorphic Art,* trans. W. J. Strachan (New York: Harry N. Abrams, 1977), 1.

39. Van Etten, *Mathematical Recreations, or a Collection of Sundrie excellent Problems of ancient & modern Philosophers, Both usefull and Recreative* (London, Printed for Rich. Hawkins, 1633), 99.

40. Lacan, *The Four Fundamental Concepts of Psycho-Analysis,* trans. Alan Sheridan (New York: W.W. Norton, 1981).

41. Antoine Arnauld and P. Nicole, *Logic, or the Art of Thinking,* ed. James Dickoff and Patricia Jones (Indianapolis: Bobbs-Merrill Library of Liberal Arts, 1964).

42. Lyons, 169.

43. John Milton, *Complete Poems and Major Prose,* ed. Merritt Y. Hughes (New York: Macmillan, 1957). All references to Milton's work are taken from this edition.

44. Irigaray, *Speculum of the Other Woman,* trans. by Gillian C. Gill (Ithaca: Cornell University Press, 1985).

45. Frederic Goldin, *The Mirror of Narcissus in the Courtly Love Lyric* (Ithaca: Cornell University Press, 1967), 4.

46. Jacques Lacan, "The Mirror Stage as Formative of the Function of the I as Revealed in Psychoanalytic Experience," *Ecrits: A Selection,* trans. Alan Sheridan (New

York: Norton Press, 1977), 2. See also John P. Muller and William J. Richardson, *Lacan and Language: A Reader's Guide to "Ecrits"* (New York: International Universities Press, 1982).

47. Jacques Lacan, *Ecrits*, trans. Alan Sheridan (New York: W.W. Norton, 1977), 2.

48. Jacques Lacan, "The Two Narcissisms," *The Seminar of Jacques Lacan. Book I: Freud's Papers on Technique 1953–54,* ed. Jacques-Alain Miller, trans. John Forrester (New York: W.W. Norton, 1988), 125.

49. Lucien Dällenbach, *The Mirror in the Text,* trans. Jeremy Whiteley with Emma Huges (Chicago: University of Chicago press, 1989), 11. Also useful on this point is Svetlana Alpers' *The Art of Describing: Dutch Art in the Seventeenth Century* (Chicago: University of Chicago Press, 1983). In her chapter 2, on Kepler's model of the eye, she writes that mirrors, maps and "eyes also can take their place alongside of art as forms of picturing so understood" (26).

50. Lyons, 167.

51. Leone Battista Alberti, *De Pictura,* Book II, section 26. *On Painting and On Sculpture,* trans. Cecil Grayson (London: Phaidon Press, 1972), 61–62.

## Epilogue

1. Shuger, *Habits of Thought in the English Renaissance: Religion, Politics and the Dominant Culture* (Berkeley: University of California Press, 1990), 9.

2. Rosalie Colie, *Paradoxia Epidemica* (Princeton: Princeton University Press, 1966), 303.

3. Wolfgang Iser, *Prospecting: From Reader Response to Literary Anthropology* (Baltimore: Johns Hopkins University Press, 1989), 265.

4. Augustine, *Cat. Rud.* 2, 3.

5. Ricoeur, in *A Ricoeur Reader,* ed. Mario J. Valdes (Toronto: Toronto University Press, 1990), 98.

6. See Gombrich, 395, and Iser, 267, on the "impact function" of literature in reader response criticism, as opposed to the "explanatory function" associated with Bacon.

7. Wlad Godich, "Introduction: Caution! Reader at Work!" in *Blindness and Insight: Essays in the Rhetoric of Contemporary Criticism* by Paul de Man (London: Methuen & Co., Ltd., 1983), xxvi.

8. Ibid., xx.

9. Paul Rabinow, "Artificiality and Enlightenment: From Sociobiology to Biosociality," *Incorporations,* ed. Jonathan Crary and Sanford Kwinter (New York: Zone Press, 1992), 234–52. See also François Dagognet, *La Maitrise du vivant* (Paris: Hachette, 1988).

10. Rabinow, 248.

11. Dagognet, 12.

# Bibliography

## Works Before 1700*

Alberti, Leone Battista. *De pictura*, Book II. *On Painting and On Sculpture,* trans. Cecil Grayson. London: Phaidon Press, 1972.

Aquinas, Thomas. *Commentary on Aristotles's "On Interpretation" (Peri Hermeneias, comm. Aquinas and Cajetan)*, trans. Jean T. Oesterle. Milwaukee: Marquette University Press, 1962.

———. *Summa Theologica,* trans. Fathers of the English Dominican Province. New York: Benziger, 1947.

Aristotle. *The "Art" of Rhetoric* (c.330 B.C.), trans. John Henry Freese. London: Loeb Classical Library, 4th repr., 1959.

———. *The Poetics,* trans. W. Hamilton Fyfe. (Based on the eleventh-century Paris manuscript and Vahlen's third edition, Leipzig, 1885). London: Loeb Classical Library, 9th repr., 1982.

———. *Prior and Posterior Analytics* (c.350–344 B.C.), trans. John Warrington. London: J.M. Dent & Sons, 1964.

Arnauld, Antoine, and Nicole, P. *Logic, or The Art of Thinking (La Logique ou l'art de penser,* Paris, 1662), ed. James Dickoff and Patricia Jones. Indianapolis: Bobbs-Merrill Library of Liberal Arts, 1964.

Augustine. *The Confessions (Confessiones,* c.397–98 A.D.), ed. R. S. Pine-Coffin. Harmondsworth, Middlesex: Penguin Books, 2nd ed., 19th repr., 1985.

———. *De catechizandis rudibus* (c.402–403 A.D.). *Opera omnia,* ed. Benedictines of St. Maur, Paris, 1836–38.

———. *De Doctrina Christiana* (c.396 A.D.), trans. D. W. Robertson, Jr. New York: Macmillan, 1st ed., 23rd repr., 1958.

———. *De Magistro (Concerning the Teacher)*, trans. George G. Leckie. London: D. Appleton-Century Company, 1938.

Bacon, Francis. *The Works of Francis Bacon,* ed. James Spedding, R. L. Ellis, and D. D. Heath, in 14 volumes. New York: Garrett Press, 1968.

———. *The Philosophical Works of Francis Bacon, Reprinted from the Texts . . . of Ellis and Spedding,* ed. John M. Robertson. New York, 1905.

---

*Where relevant, the probable date of composition is given in parentheses following the title.

———. *The Masculine Birth of Time (Temporis Partus Masculus*, 1603), *Thoughts and Conclusions (Cogitata et Visa,* 1607), *The Refutation of Philosophies (Redargutio Philosophiarum,* 1608), in *The Philosophy of Francis Bacon,* trans. Benjamin Farrington. Liverpool: Liverpool University Press, 1970.

———. *The Advancement of Learning and New Atlantis* (1605), ed. Arthur Johnston. Oxford: Clarendon Press, 1974.

Boyle, Robert. *The Works of the Honourable Robert Boyle,* ed. Thomas Birch. 6 vols. London, 1772.

———. *The Christian Virtuoso, shewing that by being addicted to Experimental Philosophy, a Man is rather Assisted, than Indisposed, to be a Good Christian.* London, 1690.

———. *Some Considerations Touching the Style of the Holy Scriptures, extracted from several parts of a Discourse concerning divers particulars belonging to the Bible written divers Years since to a Friend.* London, 1661.

Cicero. *De Oratore* (55 B.C.), trans. E. W. Sutton. 2 vols. Cambridge, Mass.: Loeb Classical Library, 1st ed., 4th repr., 1942.

Erasmus, Desiderius. *Collected Works of Erasmus,* ed. Craig R. Thompson. Toronto: University of Toronto Press, 1978.

———. *Apologia de In Principio Erat Sermo. Opera omnia,* vol. IX, ed. J. LeClerc. Leiden, 1703–1706; repr. London: Gregg Press, 1962.

Gassendi, Pierre. *Syntagma Philosophicum,* in *Opera Omnia Book I: De logicae fine.* Lyons, 1658.

Glanville, Joseph. *The Vanity of Dogmatizing: The Three Versions* (1661, 1665, 1676), ed. Stephen Medcalf. Sussex: Harvester Press, 1970.

Heraclitus. *The Art and Thought of Heraclitus,* an edition of the fragments with translation and commentary by Charles H. Kahn. Cambridge: Cambridge University Press, 1979.

Hobbes, Thomas. *Leviathan, Or The Matter, Form and Power of A Commonwealth, Ecclesiastic & Civil,* ed. W. G. Smith. Reprinted from the edition of 1651. Oxford: At the Clarendon Press, 1909. Also see *Leviathan,* ed. John Plamentz. Glasgow: William Collins Sons & Co., 1962.

———. *The Art of Rhetorick. The English Works of Thomas Hobbes of Malmesbury,* ed. Sir William Molesworth. Vol. VI. London, 1839; repr. Germany: Scientia Aalen, 1962.

———. *Elements of Philosophy. The English Works of Thomas Hobbes of Malmesbury,* ed. Sir William Molesworth. Vol I. London, 1839; repr. Germany: Scientia Aalen, 1962.

Lamy, Bernard. *The Art of Speaking: Written in French by Messieurs du Port Royal: In Pursuance of a former Treatise, Intuitled "The Art of Thinking."* Rendered into English. London, 1676.

Lever, Ralph. *The Arte of Reason, rightly termed Witcraft.* London, 1573.

Locke, John. *Essay On Human Understanding* (1689), in *The Works of John Locke,* vol. II, London 1823; repr. Germany: Scientia Verlag Aalen, 1963.

Milton, John. *Complete Shorter Poems,* ed. John Carey. Essex: Longman Group, 1971.

———. *Complete Poems and Major Prose.* Ed. Merritt Y. Hughes. New York: Macmillan, 1957.

———. *A Fuller Course in the Art of Logic,* trans. Walter J. Ong and Charles J. Ermatinger. *Complete Prose Works of John Milton.* New Haven: Yale University Press, 1982.

———. *"Paradise Lost" and Selected Poetry and Prose,* ed. Northrop Frye. New York: Holt, Rinehart & Winston, 1951.

———. *The Reason of Church Government Urged Against Prelaty* (1642). *The Prose Works of John Milton,* ed. J. A. St. John. Vol. II. London: George Ball and Sons, 1893.

Montaigne, Michel. *Essays,* trans. J.M. Cohen. London: Penguin Books, 1958.

Pascal, Blaise. *On the Geometrical Mind and the Art of Persuasion* (1657–58), in *Pascal Selections,* ed. and trans. Richard H. Popkin. New York: Macmillan, 1989.

Peter of Spain (Petrus Hispanus). *The Summulae logicales of Peter of Spain* (1246), ed. J. P. Mullally, Notre Dame: University of Notre Dame Press, 1945. See also *Petri Hispani Summulae logicales,* ed. I. M. Bochenski. Turin, 1947.

Plato. *Cratylus, Or, On the Correctness of Names,* trans. H. N. Fowler. Cambridge, Mass.: Loeb Classical Library, 1st ed., 5th repr., 1977.

———. *Gorgias,* trans. Walter Hamilton. New York: Viking Penguin, 11th repr., 1985.

———. *Phaedo,* trans. David Gallop. Oxford: Clarendon Press, 1975.

———. *Phaedrus and The Seventh and Eighth Letters* (c.411–404 B.C.), trans. Walter Hamilton. New York: Viking Penguin, 1st ed., 5th repr., 1985.

———. *The Republic* (c.375 B.C.), trans. B. Jowett. Oxford: Clarendon Press, 1888.

Quintilian, *Institutio Oratio,* trans. H. E. Butler. 4 vols. Cambridge, Mass.: The Loeb Classical Library, 1958.

Ramus, Peter. *The Logike of the Moste Excellent Philosopher P. Ramus, Martyr,* trans. Roland MacIlmaine. London, 1574.

Richardson, Alexander. *The Logicians School-Master, or a Comment upon Ramus' Logicke.* London: Printed for John Bellamie, 1629.

Shakespeare, William. *The Riverside Shakespeare.* Boston: Houghton Mifflin, 1974.

Sidney, Philip. *A Defence of Poetry,* ed. Jan Van Dorsten. (Based on the 1595 edition of the *Defence of Poesie,* London, printed for William Ponsonby). Oxford: Oxford University Press, 1966.

Sprat, Thomas. *The History of the Royal Society of London, For the Improving of Natural Knowledge.* London, 1667. Repr. in *English Science, Bacon to Newton,* ed. Brian Vickers. Cambridge: Cambridge University Press, 1987.

Van Etten, Henry. *Mathematical Recreations, or a Collection of Sundrie excellent Problems of ancient & modern Philosophers, Both usefull and Recreative.* London: Printed for Rich. Hawkins, 1633.

Ward, Seth. *Vindiciae Academiarum.* London, 1654. Repr. in *Science and Education in the Seventeenth Century: The Webster-Ward Debate,* ed. Allen G. Debus. London: MacDonald & Co., 1970.

Webster, John. *Academiarum Examen.* London, 1653. Repr. in *Science and Education in the Seventeenth Century: The Webster-Ward Debate,* ed. Allen G. Debus. London: MacDonald & Co., 1970.

Wilkins, John. "Dedicatory Epistle," *An Essay Towards a Real Character and a Philosophical Language.* London: Printed for S. Geillibrand and John Martyn, for the Royal Society, 1668.

Wilson, Thomas. *The Rule of Reason, conteinying the arte of Logique set forth in Englishe.* London: Richard Grafton, 1551.

———. *The Arte of Rhetorique, for the use of all such as are studious of eloquence.* London: Richard Grafton, 1553.

## Works After 1700

Albanese, Denise. *New Science, New World.* Durham, N.C.: Duke University Press, 1996.

Allen, D. C. "Some Theories of the Growth and Origin of Language in Milton's Age." *Philological Quarterly* 28 (1948): 5–16.

Alpers, Svetlana. *The Art of Describing: Dutch Art in the Seventeenth Century.* Chicago: University of Chicago Press, 1983.

Anderson, Judith H. *Words That Matter: Linguistic Perception in Renaissance English.* Stanford: Stanford University Press, 1996.

Ashworth, E. J. *Language and Logic in the Post Medieval Period.* Dordrecht, Holland: D. Reidel, 1974.

Auerbach, Erich. *Mimesis: The Representation of Reality in Western Literature,* trans. Willard R. Trask. Princeton: Princeton University Press, 4th repr., 1974.

Baldwin, Charles Sears. *Medieval Rhetoric and Poetic.* New York: The Macmillan Co., 1928.

Baltrusaitis, Jurgis. *Anamorphic Art,* trans. W. J. Strachan. New York: Harry N. Abrams, 1977.

Barker, Francis. "In the Wars of Truth." *Southern Review* 20 (1987): 111–25.

Barker, John. *Strange Contrarieties: Pascal in England during the Age of Reason.* Montreal: McGill University Press, 1975.

Barnard, T. C. *Cromwellian Ireland: English Government and Reform in Ireland, 1649–60.* Oxford: Oxford University Press, 1975.

Barton, Anne. *The Names of Comedy.* Toronto: University of Toronto Press, 1990.

Baxandall, Michael. *Giotto and the Orators: Humanist Observers of Painting in Italy and the Discovery of Pictorial Composition, 1350–1450.* Oxford: Clarendon Press, 1971.

Bentley, Jerry H. *Humanists and Holy Writ: New Testament Scholarship in the Renaissance.* Princeton: Princeton University Press, 1983.

Biagoli, Mario. "From Relativism to Contingentism." In *The Disunity of Science: Boundaries, Contexts, and Power.* Ed. Peter Balison and David J. Stump. Stanford: Stanford University Press, 1996.

Black, Max. "How Metaphors Work." In *On Metaphor,* ed. Sheldon Sacks. Chicago: University of Chicago Press, 1978.

Blumenberg, Hans. *The Genesis of the Copernican World.* Trans. Robert M. Wallace. Cambridge, Mass.: MIT Press, 1987.

Boyle, Marjorie O'Rourke. *Erasmus on Language and Method in Theology.* Toronto: University of Toronto Press, 1977.

Breen, Quirinus. "Giovanni Pico Della Mirandola on the Conflict of Philosophy and Rhetoric." *Journal of the History of Ideas* (1952): 384–412.

Briggs, John C. *Francis Bacon and the Rhetoric of Nature.* Cambridge, Mass.: Harvard University Press, 1989.

Calderwood, James L. *Metadrama in Shakespeare's "Henriad": Richard II to Henry V.* Berkeley: University of California Press, 1979.

———. *Shakespearean Metadrama.* Minneapolis: University of Minnesota Press, 1971.

Canny, Nicholas. *The Upstart Earl: A Study of the Social and Mental World of Richard Boyle First Earl of Cork.* Cambridge: Cambridge University Press, 1982.

Cassirer, Ernst. *An Essay on Man.* New Haven: Yale University Press, 1944.

———. *The Philosophy of Symbolic Forms.* New Haven: Yale University Press, 1953.
Cave, Terence. *The Cornucopian Text.* Oxford: Clarendon Press, 1979.
Clark, Ruth. *Strangers and Sojourners at Port Royal.* Cambridge: University Press, 1932.
Cluett, Robert. "Style, Precept, Personality: A Test Case (Thomas Sprat, 1635–1713)." *Computers and the Humanities* 5 (1970): 257–77.
Cohen, Murray. *Sensible Words: Linguistic Practice in England, 1640–1785.* Baltimore: Johns Hopkins University Press, 1977.
Cohen, Ted. "Metaphor and the Cultivation of Intimacy." In *On Metaphor,* ed. Sheldon Sacks. Chicago: University of Chicago Press, 1978.
Colie, Rosalie. *Paradoxia Epidemica.* Princeton: Princeton University Press, 1966.
Coquillette, Daniel R. *Francis Bacon,* in the series *Jurists: Profiles in Legal Theory.* Edinburgh: Edinburgh University Press, 1992.
Croll, Morris W. "'Attic Prose' in the Seventeenth Century." *Studies in Philology* 18 (April 1921): 79–128.
———. "Attic Prose: Lipsius, Montaigne, Bacon." *Schelling Anniversary Papers by His Former Students.* New York: The Century Company, 1923, 117–150.
———. "The Baroque Style in Prose." *Studies in English Philology: A Miscellany in Honor of Frederick Klaeber,* ed. Kemp Malone and Martin B. Ruud. Minneapolis: University of Minnesota Press, 1929.
———. *Style, Rhetoric, and Rhythm: Essays by Morris W. Croll.* Ed. J. Max Patrick, Robert O. Evans. Princeton, N.J.: Princeton University Press, 1966.
Culler, Jonathan. *Structuralist Poetics: Structuralism, Linguistics and the Study of Literature.* London: Routledge & Kegan Paul, 1975.
Curtius, Ernst. *European Literature and the Latin Middle Ages,* trans. W. Trask. New York: Pantheon Books, 1953.
Dagognet, François. *La Maitrise du vivant.* Paris: Hachette, 1988.
Dällenbach, Lucien. *The Mirror in the Text.* Trans. Jeremy Whiteley with Emma Hughes. Chicago: University of Chicago Press, 1989.
Daniel, Stephen H. "Myth and the Grammar of Discovery in Francis Bacon." *Philosophy and Rhetoric* 15 (1982): 219–37.
Davidson, Donald. "What Metaphors Mean." In *On Metaphor,* ed. Sheldon Sacks. Chicago: University of Chicago Press, 1978.
Dear, Peter. "Totius in Verba: Rhetoric and Authority in the Early Royal Society." *Isis* 76 (1985): 145–61.
de Grazia, Margreta. "The Secularization of Language in the Seventeenth Century." *Journal of the History of Ideas* 41 (1980): 319–29.
de Man, Paul. *Blindness and Insight: Essays in the Rhetoric of Contemporary Criticism,* ed. Wlad Godzich. London: Methuen, 2nd ed., 1983.
———. "The Epistemology of Metaphor." In *On Metaphor,* ed. Sheldon Sacks. Chicago: University of Chicago Press, 1978.
———. "Pascal's Allegory." *Allegory and Representation,* ed. Stephen Jay Greenblatt. Baltimore: Johns Hopkins University Press, 1981.
Dubois, Paige. "Subjected Bodies, Science, and the State: Francis Bacon, Torturer." In *Body Politics: Disease, Desire, and the Family,* ed. Michael Ryan and Avery Gordon. Boulder, Col.: Westview Press, 1994.
Eagleton, Terry. *Literary Theory: An Introduction.* Oxford: Basil Blackwell, 1986.

Edgerton, Samuel Y. *The Renaissance Rediscovery of Linear Perspective*. New York: Basic Books, 1975.

Else, G. R. "'Imitation' in the Fifth Century." *Classical Philology* 53 (1958): 73–90.

Elsky, Martin. *Authorizing Words: Speech, Writing, and Print in the English Renaissance*. Ithaca: Cornell University Press, 1989.

Epstein, Joel J. *Francis Bacon: A Political Biography*. Athens: Ohio University Press, 1977.

Fantazzi, Charles. *Juan Luis Vives' in Pseudodialecticos: A Critical Edition*. Dordrecht: 1979.

Ferguson, Margaret W. *Trials of Desire: Renaissance Defenses of Poetry*. New Haven: Yale University Press, 1983.

Fish, Stanley. "Literature in the Reader: Affective Stylistics." *New Literary History* 2 (1970).

———. "Reasons that Imply Themselves: Imagery, Argument, and the Reader in Milton's *Reason of Church Government*." *Seventeenth-Century Imagery: Essays on Uses of Figurative Language from Donne to Farquhar*, ed. Earl Miner. Berkeley: University of California Press, 1971.

———. *Self-Consuming Artifacts*. Berkeley: University of California Press, 1972.

Foucault, Michel. *The Order of Things* (*Les mots et les choses*. Paris: Editions Gallimard, 1966) London: Tavistock/Routledge, 1970.

Fraser, Russell. *The Language of Adam: On the Limits and Systems of Discourse*. New York: Columbia University Press, 1977.

Freedberg, David. *The Power of Images*. Chicago: University of Chicago Press, 1989.

Freud, Sigmund. "Instincts and Their Vicissitudes." In *On Metaphyschology: The Theory of Psychoanalysis*, trans. James Strachey, Penguin Freud Library, vol. 2. London: Penguin Books, 1984.

Funkenstein, A. *Theology and the Scientific Imagination*. Princeton, N.J.: Princeton University Press, 1986.

Gilman, Ernest B. *The Curious Perspective: Literary and Pictorial Wit in the Seventeenth Century*. New Haven: Yale University Press, 1978.

Godzich, Wlad. "Introduction: Caution! Reader at Work." In Paul de Man, *Blindness and Insight: Essays in the Rhetoric of Contemporary Criticism*. London: Methuen, 1983.

Goldberg, Jonathan. *Voice Terminal Echo, Postmodernism and English Renaissance Texts*. New York: Methuen, 1986.

Goldin, Frederic. *The Mirror of Narcissus in the Courtly Love Lyric*. Ithaca: Cornell University Press, 1967.

Gombrich, E. H. *Art and Illusion: A Study in the Psychology of Pictorial Representation*. Bollingen Series XXXV. Princeton: Princeton University Press, 2nd ed., 9th repr., 1989.

Gray, Hanna H. "Renaissance Humanism: The Pursuit of Eloquence." *Journal of the History of Ideas* (1963): 497–514.

Greene, Thomas M. *The Light in Troy: Imitation and Discovery in Renaissance Poetry*. New Haven: Yale University Press, 1982.

Guillen, Claudio. "On the Concept and Metaphor of Perspective." In *Comparatists at Work*, ed. Stephen G. Nichols, Jr., Richard B. Vowles. Waltham, Mass.: Blaisdell Publishing Company, 1968. *Papers on French Seventeenth-Century Literature* 10 (1983): 177–97.

Habermas, Jürgen. *Toward a Rational Society: Student Protest, Science, and Politics.* Boston: Beacon Press, 1970.

Hall, Joan Wylie. "Bacon's Triple Curative: The 1597 *Essayes, Meditations,* and *Places.*" *Papers on Language & Literature* 21 (1985): 345–58.

Haraway, Donna. "A Manifesto for Cyborgs: Science, Technology, and Socialists Feminism in the 1980s." In *Socialist Review* 80 (March–April 1985): 65–108.

Hardison, O. B. "The Two Voices of Sidney's *Apology for Poetry.*" *English Literary Renaissance* 2 (1972): 83–99.

Harwood, John T. "Introduction." In *The Rhetorics of Thomas Hobbes and Bernard Lamy.* Carbondale: Southern Illinois University Press, 1986.

———. "Science Writing and Writing Science: Boyle and Rhetorical Theory." In *Boyle Reconsidered.* Ed. Michael Hunter. Cambridge: Cambridge University Press, 1994.

Hattaway, Michael. "Bacon and 'Knowledge Broken': Limits for the Scientific Method." *Journal of the History of Ideas* 39 (1978): 183–97.

Havelock, Eric. *Preface to Plato.* Oxford: Basil Blackwell, 1963.

Hawkins, Peter, ed. *Ineffability: Naming the Unnameable from Dante to Beckett.* New York: AMS Press, 1984.

Herrick, Marvin T. "The Early History of Aristotle's *Rhetoric* in England." *Philological Quarterly* 5 (1926): 242–57.

Hoopes, Robert. *Right Reason in the English Renaissance.* Cambridge, Mass.: Harvard University Press, 1962.

Horkheimer, Max, and Theodor W. Adorno. *Dialectic of Enlightenment.* New York: Continuum Press, 1993.

Hovey, Kenneth Alan. "'*Mountaigny* Saith Prettily': Bacon's French and the Essay." *PMLA* 106 (1991): 71–82.

Howell, A. C. "*Res et Verba*: Words and Things." *English Literary History* 13 (1946): 131–42.

Howell, W. S. *Logic and Rhetoric in England, 1500–1700.* Princeton: Princeton University Press, 1956.

Hulse, Clark. "Spenser, Bacon and the Myth of Power." In *The Historical Renaissance: New Essays on Tudor and Stuart Literature and Culture,* ed. Heather Dubrow and Richard Strier. Chicago: University of Chicago Press, 1988.

Hunter, J. Paul. "Robert Boyle and the Epistemology of the Novel." *Eighteenth-Century Fiction* 2 (July 1990): 275–82.

———. *Before Novels: The Cultural Contexts of Eighteenth-Century English Fiction.* New York: W.W. Norton, 1990.

Hunter, Michael. *Robert Boyle Reconsidered.* Cambridge: Cambridge University Press, 1994.

———. *Science and Society in Restoration England.* Cambridge: Cambridge University Press, 1981.

———. *Establishing the New Science: The Experience of the Early Royal Society.* Woodbridge, Suffolk [England]: Boydell Press, 1989.

Hussey, Edward. "Epistemology and Meaning in Heraclitus." In *Language and Logos: Studies in Ancient Greek Philosophy Presented to G. E. L. Owen,* ed. Malcolm Schofield and Martha C. Nussbaum. Cambridge: Cambridge University Press, 1982.

Huxley, Aldous. *Literature and Science*. London: Chatto & Windus, 1970.

Ijsseling, Samuel. *Rhetoric and Philosophy in Conflict*. The Hague: M. Nijhoff, 1976.

Irigaray, Luce. *Speculum of the Other Woman*, translated by Gillian C. Gill. Ithaca: Cornell University Press, 1985.

Iser, Wolfgang. *Prospecting: From Reader Response to Literary Anthropology*. Baltimore: Johns Hopkins University Press, 1989.

Jacob, J. R. *Robert Boyle and the English Revolution: A Study in Social and Intellectual Change*. New York: B. Franklin, 1977.

Jacob, J. R., and Margaret C. Jacob. "The Anglican Origins of Modern Science: The Metaphysical Foundations of the Whig Constitution." *Isis* 71 (1980): 251–67.

Jacob, Margaret C. *The Cultural Meaning of the Scientific Revolution*. Philadelphia: Temple University Press, 1988.

Jameson, Frederic. *The Prison-House of Language: A Critical Account of Structuralism and Russian Formalism*. Princeton: Princeton University Press, 1972.

Jardine, Lisa. *Francis Bacon: Discovery and the Art of Discourse*. Cambridge: Cambridge University Press, 1974.

Jarrott, C. A. L. "Erasmus' *In Principio Erat Sermo:* A Controversial Translation." *Studies in Philology* 61 (1964): 35–40.

Johnston, David. *The Rhetoric of Leviathan: Thomas Hobbes and the Politics of Cultural Transformation*. Princeton: Princeton University Press, 1986.

Jones, R. F. *The Triumph of the English Language*. Stanford: Stanford University Press, 1953.

———. "The Background of *The Battle of the Books*." *Washington University Studies* 7 Humanistic Series II (1920).

———. "Science and Criticism in the Neo-Classical Age of English Literature." *Journal of the History of Ideas*, Vol. I (1940).

———."Science and English Prose Style in the Third Quarter of the Seventeenth Century." *Publications of the Modern Language Association* 45 (1930).

———. "Science and Language in England of the Mid-Seventeenth Century." *The Journal of English and Germanic Philology*, Vol. XXXI (1932).

Josipovici, Gabriel. *The World and the Book*. 2nd ed. London: Macmillan Press, 1979.

Kahn, Charles H. *The Art and Thought of Heraclitus,* an edition of the fragments with translation and commentary. Cambridge: Cambridge University Press, 1979.

Kahn, Victoria. "Humanism and the Resistance to Theory." In *Literary Theory/Renaissance Texts,* ed. Patricia Parker and David Quint. Baltimore: Johns Hopkins University Press, 1986.

———. *Rhetoric, Prudence, and Skepticism in the Renaissance*. Ithaca: Cornell University Press, 1985.

Katz, David S. "The Language of Adam in Seventeenth-Century England." *History and Imagination: Essays in Honor of H. R. Trevor-Roper,* ed. Hugh Lloyd-Jones. New York: Holmes and Meier, 1982.

Kearney, Hugh. *Scholars and Gentlemen: Universities and Society in Pre-Industrial Britain, 1500–1700.* London: Faber and Faber, 1970.

Keeble, N. H. *The Literary Culture of Noncomformity in Later Seventeenth-Century England*. Leicester: Leicester University Press, 1987.

Keller, Evelyn Fox. "Nature, Nurture, and the Human Genome Project." In *The Code of Codes: Scientific and Social Issues in the Human Genome Project,* ed. Daniel J. Kevles and Leroy Hood. Cambridge, Mass.: Harvard University Press, 1992.

———. *Secrets of Life, Secrets of Death: Essays on Language, Gender and Science.* New York: Routledge, 1992.

———. "The Dilemma of Scientific Subjectivity in Postvital Culture." In *The Disunity of Science: Boundaries, Contexts, and Power,* ed. Peter Galison and David J. Stump. Stanford: Stanford University Press, 1996.

Kennedy, George A. *Classical Rhetoric and Its Christian and Secular Tradition.* Chapel Hill: University of North Carolina Press, 1980.

Kennedy, William J. *Rhetorical Norms in Renaissance Literature.* New Haven: Yale University Press, 1978.

King, Donald B., and H. David Rix, trans. and eds. Erasmus, Desiderius. *On Copia of Words and Ideas.* Milwaukee: Marquette University Press, 1963.

Kneale, William, and Martha. *The Development of Logic.* Oxford: Clarendon Press, 1984.

Kristeller, Paul Oscar. *Renaissance Thought and Its Sources,* ed. Michael Mooney. New York: Columbia University Press, 1979.

Krook, Dorothea. "Two Baconians: Robert Boyle and Joseph Glanville." *Huntington Library Quarterly* 18 (1955): 261–78.

Lacan, Jacques. *The Four Fundamental Concepts of Psycho-Analysis.* Trans. Alan Sheridan. New York: W.W. Norton, 1981.

———. "The Mirror Stage as Formative of the Function of the 'I' as Revealed in Pscholanalytical Experience." *Ecrits: A Selection,* trans. Alan Sheridan. New York: W.W. Norton, 1977.

———. "The Two Narcissisms," *The Seminar of Jacques Lacan. Book I: Freud's Papers on Technique 1953–54,* ed. Jacques-Alain Miller, trans. John Forrester. New York: W.W. Norton, 1988.

Langbein, John H. *Prosecuting Crime in the Renaissance.* Cambridge, Mass.: Harvard University Press, 1974.

Lanham, Richard A. *The Motives of Eloquence: Literary Rhetoric in the Renaissance.* New Haven: Yale University Press, 1976.

Latour, Bruno. *We Have Never Been Modern.* Trans. Catherine Porter. Cambridge, Mass.: Harvard University Press, 1993.

Lechner, Sister Joan Marie, O.S.U. *Renaissance Concepts of the Commonplaces.* New York: Pageant Press, 1962.

Leeman, Fred. *Hidden Images: Games of Perception, Anamorphic Art, Illusion,* trans. Ellyn Childs Allison and Margaret L. Kaplan. New York: Harry N. Abrams, 1976.

Lemmi, Charles. *The Classical Deities in Bacon: A Study in Mythological Symbolism.* Baltimore: Johns Hopkins University Press, 1933.

Levack, Brian P. *The Civil Lawyers in England, 1603–1641.* Oxford: Clarendon Press, 1973.

Levy, F. J. "Francis Bacon and the Style of Politics." *English Literary Renaissance* 16 (1986): 102–22.

Lewalski, Barbara Kiefer. *Protestant Poetics and the Seventeenth-Century Religious Lyric.* Princeton: Princeton University Press, 1979.

Lindberg, David. C. *John Pecham and the Science of Optics.* Madison: University of Wisconsin Press, 1970.

Long, Pamela. "Humanism and Science." In *Renaissance Humanism: Foundations, Forms, and Legacy,* vol. 3: *Humanism and the Disciplines,* ed. Albert Rabil, Jr. Philadelphia: University of Pennsylvania Press, 1988.

Lovejoy, Arthur O. *The Great Chain of Being: A Study of the History of an Idea.* Cambridge, Mass.: Harvard University Press, 1936.

Lyons, John D., and Stephen G. Nichols, eds. *Mimesis: From Mirror to Method, Augustine to Descartes.* Hanover, N.H.: University Press of New England, 1982.

Lyons, John D. "Speaking in Pictures, Speaking of Pictures: Problems of Representation in the Seventeenth Century." In *Mimesis: From Mirror to Method, Augustine to Descartes,* ed. John D. Lyons and Stephen G. Nichols. Hanover, N.H.: University Press of New England, 1982.

MacCaffrey, Isabel. *Paradise Lost as "Myth."* Cambridge, Mass.: Harvard University Press, 1959.

McCanles, Michael. *Dialectical Criticism and Renaissance Literature.* Berkeley: University of California Press, 1975.

———. "From Derrida to Bacon and Beyond." In *Francis Bacon's Legacy of Texts,* ed. William A. Sessions. New York: AMS, 1990.

MacIntosh, J. J. "Perception and Imagination in Descartes, Boyle and Hooke." *Canadian Journal of Philosophy* 13 (1983): 327–52.

Marcuse, Herbert. "Remarks on a Redefinition of Culture." In *Science and Culture: A Study of Cohesive and Disjunctive Forces,* ed. Gerald Holton. Boston: Houghton Mifflin, 1965.

Markley, Robert. "Robert Boyle on Language." *Studies in Eighteenth-Century Culture* 14 (1985): 159–71.

———. "Objectivity as Ideology: Boyle, Newton and the Language of Science." *Genre* 16 (1983): 355–72.

Martin, Julian. *Francis Bacon, the State, and the Reform of Natural Philosophy.* Cambridge: Cambridge University Press, 1992.

Martin-Trigona, Helen Vasiliou. "Logical Proof and Imaginative Reason in Selected Speeches of Francis Bacon." Diss., University of Illinois, 1967.

Mathie, William. "Reason and Rhetoric in Hobbes's *Leviathan.*" *Interpretation* 14 (1986):281-98.

Mazzeo, Joseph. "St. Augustine's Rhetoric of Silence." *Journal of the History of Ideas* 23 (1962): 175–96.

Merchant, Carolyn. *The Death of Nature: Women, Ecology, and the Scientific Revolution.* San Francisco: Harper & Row, 1980.

Miall, David S. "Metaphor as a Thought-Process." *Journal of Aesthetics and Art Criticism* 38 (1979): 21–28.

Moore, F. C. T. "On Taking Metaphor Literally." In *Metaphor: Problems and Perspective,* ed. David Miall. New York: Harvester and Humanities Presses, 1982.

Morgan, John. "Puritanism and Science: A Reinterpretation." In *Historical Journal* 20 (1979): 535–60.

Mullally, J. P. *The Summulae Logicales of Peter of Spain.* Notre Dame, IN: University of Notre Dame Press, 1945.

Muller, John P., and William J. Richardson *Lacan and Language: A Reader's Guide to "Ecrits."* New York: International Universities Press, 1982.

Mulligan, Lotte. "'Reason,' 'Right Reason,' and 'Revelation' in Mid-Seventeenth Century England." *Occult and Scientific Mentalities in the Renaissance,* ed. Brian Vickers. Cambridge: Cambridge University Press, 1984.

Murphy, James J., ed. *Renaissance Eloquence: Studies in the Theory and Practice of Renaissance Rhetoric.* Berkeley: University of California Press, 1983.

Norbrook, David. *Poetry and Politics in the English Renaissance.* London: Routledge & Kegan Paul, 1984.

Nuttall, A. D. *A New Mimesis: Shakespeare and the Representation of Reality.* London: Methuen, 1983.

Ong, Walter, S. J. *Ramus, Method, and the Decay of Dialogue, From the Decay of Dialogue to the Art of Reason.* Cambridge, Mass.: Harvard University Press, 1958.

Oster, Malcolm. "Virtue, Providence and Political Neutralism: Boyle and Interregnum Politics." In *Boyle Reconsidered,* ed. Michael Hunter. Cambridge: Cambridge University Press, 1994.

Padley, G. A. *Grammatical Theory in Western Europe, 1500-1700: Trends in Vernacular Grammar.* Cambridge: Cambridge University Press, 1985.

Pagden, Anthony, ed. *The Language of Political Theory in Early-Modern Europe.* Cambridge: Cambridge University Press, 1987.

Pagel, Walter. *Paracelsus: An Introduction to Philosophical Medicine in the Era of the Renaissance.* Basel and New York: S. Karger, 1958.

Paivio, Allan. "Psychological Processes in the Comprehension of Metaphor." In *Metaphor and Thought,* ed. Andrew Ortony. Cambridge: Cambridge University Press, 1979.

Paradis, James. "Montaigne, Boyle, and the Essay of Experience." In *One Culture: Essays in Science and Literature,* ed. George Levine. Madison: University of Wisconsin Press, 1987.

Patterson, Annabel. *Fables of Power: Aesopian Writing and Political History.* Durham, N.C.: Duke University Press, 1991.

Peck, Linda Levy. *Court Patronage and Corruption in Early Stuart England.* Boston: Unwin Hyman, 1990.

Pérez-Ramos, Antonio. *Francis Bacon's Idea of Science and the Maker's Knowledge Tradition.* Oxford: Clarendon Press, 1988.

———. "Bacon's Legacy." In *The Cambridge Companion to Bacon,* ed. Markku Peltonen. Cambridge: Cambridge University Press, 1996.

Pooley, Roger. "Language and Loyalty: Plain Style at the Restoration." *Literature and History* 6 (1980): 2–18.

Prest, Wilfred R. *The Inns of Court Under Elizabeth I and the Early Stuarts.* Totowa, N.J.: Rowan and Littlefield, 1972.

Preus, Mary C. *Eloquence and Ignorance in Augustine's "On the Nature and Origin of the Soul."* Atlanta, Ga.: Scholars Press, 1985.

Prickett, Stephen. *Words and The Word: Language, Poetics, and Biblical Interpretation.* Cambridge: Cambridge University Press, 1986.

Quilligan, Maureen. *The Language of Allegory: Defining the Genre.* Ithaca: Cornell University Press, 1979.

Quint, David. *Origin and Originality in Renaissance Literature: Versions of the Source.* New Haven: Yale University Press, 1983.

Rabinow, Paul. "Artificiality and Enlightenment: From Sociobiology to Biosociality." In *Incorporations,* ed. Jonathan Crary and Sanford Kwinter. New York: Zone Press, 1992.

Rees, Graham. "Francis Bacon's Semi-Paracelsian Cosmology." *Ambix* 22 (1975): 81–101.

———. "The Fate of Francis Bacon's Cosmology in the Seventeenth Century." *Ambix* 24 (1977): 27–38.

Reiss, Timothy. *The Discourse of Modernism.* Ithaca: Cornell University Press, 1982.

———. "Power, Poetry, and the Resemblance of Nature." In *Mimesis: From Mirror to Method, Augustine to Descartes,* ed. John D. Lyons and Stephen G. Nichols. Hanover, N.H.: University Press of New England, 1982.

Reynolds, L. D., and N. G. Wilson. *Scribes and Scholars: A Guide to the Transmission of Greek and Latin Literature.* 2nd ed., 3rd repr. Oxford: Clarendon Press, 1984.

Richardson, W. C. *A History of the Inns of Court: With Special Reference to the Period of the Renaissance.* Baton Rouge, LA: Claitor's Pub. Division, 1975.

Ricoeur, Paul. "The Metaphorical Process as Cognition, Imagination, and Feeling." In *On Metaphor,* ed. Sheldon Sacks. Chicago: University of Chicago Press, 1978.

———. *The Rule of Metaphor,* trans. Robert Czerny. London: Routledge & Kegan Paul, 1978.

———. *A Ricoeur Reader: Reflection and Imagination,* ed. Mario J. Valdes. Toronto: University of Toronto Press, 1991.

Rosenmeyer, Thomas G. "Gorgias, Aeschylus, and *Apate.*" *American Journal of Philology* 76 (1955): 225–60.

Rossi, Paolo. *Francis Bacon: From Magic to Science.* Trans. Sacha Rabinovitch. Chicago: University of Chicago Press, 1968.

———. *Philosophy, Technology, and the Arts in the Early-Modern Era.* New York: Harper & Row, 1970.

Sabra, A. I. *Theories of Light, from Descartes to Newton.* New York: American Elsevier, 1967.

Sacksteder, William. "Hobbes: Philosophical and Rhetorical Artifice." *Philosophy and Rhetoric* 17 (1984): 30–46.

Salmon, Vivian. *The Study of Language in Seventeenth-Century England.* Amsterdam Studies in the Theory and History of Linguistic Science, ed. E. F. K. Koerner, Series 3, XVII. Amsterdam: J. Benjamins, 1979.

———. *The Works of Francis Lodwick: A Study of his Writings in the Intellectual Context of the Seventeenth Century.* London: Longman, 1972.

Sargent, Rose-Mary. "Scientific Experiment and Legal Expertise: The Way of Experience in Seventeenth-Century England." *Studies in History and Philosophy of Science,* 20:1 (March 1989): 28.

———. "Learning From Experience: Boyle's Construction of an Experimental Philosophy. In *Boyle Reconsidered,* ed. Michael Hunter. Cambridge: Cambridge University Press, 1994.

———. "Robert Boyle's Baconian Inheritance: A Response to Laudan's Cartesian Thesis." *Studies in the History and Philosophy of Science* 17 (1986): 469–86.

———. *The Diffident Naturalist: Robert Boyle and the Philosophy of Experiment.* Chicago: University of Chicago Press, 1995.

Saussure, Ferdinand de. *Course in General Linguistics,* ed. Bally, Sechehaye, and Riedlinger, trans. Wade Baskin. New York: Fontana, 1959.

Schmitt, Charles B. *The Cambridge History of Renaissance Philosophy.* Cambridge: Cambridge University Press, 1988.

Schofield, Malcolm, and Martha Craven Nussbaum, eds. *Language and Logos: Studies in Ancient Greek Philosophy Presented to G. E. L. Owen.* Cambridge: Cambridge University Press, 1982.

Schwartz, Regina. *Remembering and Repeating: Biblical Creation in Paradise Lost.* Cambridge: Cambridge University Press, 1988.

———. "The Toad at Eve's Ear: From Identification to Identity." In *Literary Milton: Text, Pretext, Context,* ed. Diana Trevino Benet and Michael Lieb. Pittsburgh: Duquesne University Press, 1994.

———. "Through the Optic Glass: Voyeurism and *Paradise Lost.*" In *Desire in the Renaissance: Psychoanalysis and Literature,* ed. Valeria Finucci and Regina Schwartz. Princeton: Princeton University Press, 1994.

Sessions, William. "Bacon, Herbert and an Image in Chalk." *'Too Rich to Clothe the Sunne': Essays on George Herbert.* Ed. Claude J. Summers, Ted-Larry Pebworth. Pittsburgh: University of Pittsburgh Press, 1980.

———, ed. *Francis Bacon's Legacy of Texts: "The Art of Discovery Grows with Discovery."* New York: AMS Press, 1990.

Shapin, Steven, and Simon Schaffer. *Leviathan and the Air-Pump: Hobbes, Boyle, and the Experimental Life.* Princeton: Princeton University Press, 1985.

Shapiro, Barbara J. *Probability and Certainty in Seventeenth-Century England: A Study of the Relationships Between Natural Science, Religion, History, Law, and Literature.* Princeton: Princeton University Press, 1983.

Sharpe, Kevin. *Politics and Ideas in Early Stuart England.* London: Pinter Publishers, 1989.

Shirley, John W., and F. David Hoeniger. *Science and the Arts in the Renaissance.* Washington, D.C.: Folger Shakespeare Library, 1985.

Shuger, Debora K. *Habits of Thought in the English Renaissance: Religion, Politics and the Dominant Culture.* Berkeley: University of California Press, 1990.

———. *Sacred Rhetoric: The Christian Grand Style in The English Renaissance.* Princeton: Princeton University Press, 1988.

Slaughter, M. M. *Universal Languages and Scientific Taxonomy in the Seventeenth Century.* Cambridge: Cambridge University Press, 1982.

Sloan, Thomas O. *Donne, Milton and the End of Humanist Rhetoric.* Berkeley: University of California Press, 1985.

———. "The Crossing of Rhetoric and Poetry in the English Renaissance." In *The Rhetoric of Renaissance Poetry From Wyatt to Milton,* ed. Thomas O. Sloan and Raymond B. Waddington. Berkeley: University of California Press, 1974.

———. "Rhetoric and Meditation: Three Case Studies." *Journal of Medieval and Renaissance Studies* 1 (1971): 45–58.

Smuts, R. Malcolm. *Court Culture and the Origins of a Royalist Tradition in Early Stuart England.* Philadelphia: University of Pennsylvania Press, 1987.

Steadman, John M. *The Hill and the Labyrinth: Discourse and Certitude in Milton and His Near-Contemporaries.* Berkeley: University of California Press, 1984.

Steiner, George. *Real Presences.* London: Faber and Faber, 1989.
Stephens, James. *Francis Bacon and the Style of Science.* Chicago: University of Chicago Press, 1975.
Todd, Margo. *Christian Humanism and the Puritan Social Order.* Cambridge: Cambridge University Press, 1987.
Topliss, Patricia. *The Rhetoric of Pascal.* Leicester: Leicester University Press, 1966.
Tourangeau, Roger. "Metaphor and Cognitive Structure." In *Metaphor: Problems and Perspective,* ed. David Miall. New York: Harvester and Humanities Presses, 1982.
Tousdale, Marion. *Shakespeare and the Rhetoricians.* Chapel Hill: University of North Carolina Press, 1982.
Trinkhaus, Charles E. *In Our Image and Likeness: Humanity and Divinity in Italian Humanist Thought.* In 2 vols. Chicago: University of Chicago Press, 1970.
Tuve, Rosamond. *Elizabethan and Metaphysical Imagery: Renaissance Poetic and Twentieth-Century Critics.* Chicago: University of Chicago Press, 1947.
Vance, Eugene. *Mervelous Signals: Poetics and Sign Theory in the Middle Ages.* Lincoln: University of Nebraska Press, 1986.
Vickers, Brian, ed. *English Science, Bacon to Newton.* Cambridge: Cambridge University Press, 1987.
———. *Essential Articles for the Study of Francis Bacon.* London: Sidgwick and Jackson, 1972.
——— *Francis Bacon and Renaissance Prose.* Cambridge: Cambridge University Press, 1968.
———. "The Royal Society and English Prose Style." *Rhetoric and the Pursuit of Truth,* ed. Thomas F. Wright. Los Angeles: University of California Press, 1980.
Warhaft, Sidney. "The Providential Order in Bacon's New Philosophy." *Studies in the Literary Imagination* 4 (1971): 49–64.
Waswo, Richard. *Language and Meaning in the Renaissance.* Princeton: Princeton University Press, 1987.
Webster, Charles. *The Great Instauration: Science, Medicine and Reform, 1626–1660.* New York: Holmes & Meier, 1976.
Whitman, Jon. *Allegory: The Dynamics of an Ancient and Medieval Technique.* Cambridge, Mass.: Harvard University Press, 1987.
Whitney, Charles. *Francis Bacon and Modernity.* New Haven: Yale University Press, 1986.
Wilson, Harold S. "Some Meanings of 'Nature' in Renaissance Literary Theory." *Journal of the History of Ideas* 11 (1942): 430–48.
Wojcik, Jan W. *Robert Boyle and the Limits of Reason.* Cambridge: Cambridge University Press, 1997.
Wormald, B. H. G. *Francis Bacon: History, Politics and Science, 1561–1626.* Cambridge: Cambridge University Press, 1993.
Young, Bruce W. "Thomas Hobbes Versus the Poets: Form, Expression, and Metaphor in Early Seventeenth-Century Poetry." *Encyclia: The Journal of the Utah Academy of Sciences, Arts, and Letters* 63 (1986): 151–62.
Youngren, William H. "Generality, Science and Poetic Language in the Restoration." *English Literary History* 35 (1968): 158–88.

# Index

Adorno, Theodor, 14–15
*Advancement of Learning* (Bacon), 13, 78, 86–88; *experientia literata* in, 81–85, 100, 170 (see also under *De dignitate et augmentis, Instauratio magna, Redargutio philosophiarum*); on fables, 80, 88; influences on, 29, 98
Agricola, Rudolph, 61; *De inventione dialectica*, 63
Alberti, Leone Battista, 137
Allegory: and Bacon, 104–6, 108, 111; in *De sapientia veterum*, 95–99, 106, 107, 110–12; in the development of language, 103–4, 106; in Hobbes and *Leviathan*, 117–19, in the pre-Socratics, 104–5; and the symbolic, 101–3, 108. *See also under* Bacon; *De sapientia veterum*; Hobbes; Pre-Socratics
Alpers, Svetlana, 190 n. 49
Anamorphism: in Baroque art, 138–39; and feminism, 156, 164–65; in Irigaray, 156, 164; in *Paradise Lost* (and Eve), 161, 164; in *Paradise Regained* (and Satan) 154–56
Aphorisms: in *Advancement of Learning*, 170; in *De dignitate et augmentis*, 82–83, 87–88; in *Essays*, 181 n. 18. *See also under* Bacon
Aquinas, Saint Thomas: commentary on Aristotle, 57–58; on scientific demonstration, 58–60 (*see also under* Science)
*Areopagitica* (Milton), 150
Aristotle: and Aquinas, 57–58; and Bacon, 79, 90, 104–5. Works of: *Categories*, 55; *De interpretatione*, 55–56; *Prior & Posterior Analytics*, 56; *Topics*, 55

Arnauld, Antoine, 154
*Art and Illusion* (Gombrich), 29, 48
*Art of Logic* (Milton), 140–42. *See* Ramism
*Arte of Rhetorique, The* (Wilson), 65–67
Attic prose, 85-86, 94; 180 n. 15; 181 n. 28. *See also under* Croll
Augustine, Saint: and Bacon, 92–97; on sign systems, 19–21, 92–94, 169. Works of: *Confessions*, 92; *De doctrina christiana*, 20–21, 92; *De magistro*, 20–21, 92

Bacon, Francis, 15, 31, 76, 77, 89, 91–92; allegory, use of 104–6, 108, 111; aphorisms in, 82–83, 87–88, 170, 181 n. 18; on Aristotle, 79, 90, 104–5; and Augustine, 92–97; and Erasmus, 89, 91–92; fables, use of, 16, 22–23, 79–81, 84, 88, 91–92, 95–99, 101, 107, 110–12, 171, 183 n. 28; hieroglyphics, use of, 82–83, 96, 108, 131; methods of knowledge in, 84–88; and Milton, 25, 144–45; *mimesis* in, 49, 90–91, 101; on the pre-Socratics, 78–80, 96, 104–5, 169; on *res et verba*, 19, 97, 168, 170; and Spenser, 74–75, 107, 112; and Wilkins, 130–31; and Wilson, 66. Works of: *Advancement of Learning*, 13, 29, 78, 80, 81–85, 86–88, 98, 100, 170; *Cogitata et visa*, 78; *De dignitate & augmentis scientiarum*, 2, 22, 31, 81–85, 86–88, 96, 104; *De sapientia veterum*, 16, 22–23, 78, 79–81, 84, 88, 95–99, 106, 107, 110–12, 171, 183 n. 28; *Essays*, 73–74, 76, 77, 181 n. 18; *Instauratio magna*, 22, 76, 98, 130, 171; *Novum organon*, 19, 76, 78, 88–89, 131,

Bacon, Francis, (*continued*)
169; *Redargutio philosophiarum*, 78, 89–90; *Temporis partus masculus*, 78, 80
Baconian science: definition of, 13–14; historiography in, 54 (*see also under* Science); influence on Royal Society, 113; language, use of, 25; sources in, 15
Baltrusaitis, Jurgis, 153
Baxandall, Michael, 189 n. 37
Beale, John: letter to Boyle, 133, 186 n. 20
Black, Max, 42
Blumenberg, Hans, 54–55
Boyle, Robert, 13, 131–36; analogical reasoning in, 23, 131–32 (*see also under* Reason); civic humanism, 24, 47, 133, 134–36, 186 n. 20; dispute with Hobbes, 17, 117, 123, 132, 133–34; and Erasmus, 133; family history, 135–36; Hartlib Circle and the Invisible College, 134–36, 184 n. 3; language theory in, 133–34. Works of: *Certain Physiological Essays*, 132–33; *Christian Virtuoso*, 113, 131–32, 136, 169; *Some Considerations about Reason and Religion*, 134, 186 n. 18; *Things Said to Transcend Reason*, 134
Briggs, John C., 81, 99

Calderwood, James, 106–7
Canny, Nicholas, 135–36
Cassirer, Ernst, 103–4, 106, 188 n. 22
*Categories* (Aristotle), 55
Cave, Terence, 63–64
*Certain Physiological Essays* (Boyle), 132–33
*Christian Virtuoso* (Boyle), 113, 131–32, 136, 169
Cicero, 31; speech-act theory, 45–46, 105–6
Civic humanism: 24, 47, 133, 134–36, 186 n. 20. *See also under* Boyle; Spenser
*Cogitata et visa* (Bacon), 78
Colie, Rosalie, 168
*Confessions* (Augustine), 92
Croll, Morris: "attic" prose, 85–86, 94, 180 n. 15, 181 n. 28

Culler, Jonathan, 44–45, 176 n. 25

Dagognet, François, 172
Dällenbach, Lucien, 159
Daniel, Stephen, 104–6, 182 n. 16, 183 n. 26
Davidson, Donald, 183 n. 29
*De copia* (Erasmus), 63–65
*De dignitate et augmentis scientiarum* (Bacon), 2, 31, 86–88; on disciplinary knowledge; 22, 81; *experientia literata* in, 81–85, 96, 104 (see also under *Advancement of Learning, Instauratio magna, Redargutio philosophiarum*)
*De doctrina christiana* (Augustine), 20–21, 92
*Defence of Poetry* (Sidney): on the body politic, 70–73; on Plato, 67–68; poetics in, 51, representation in, 68–69
de Grazia, Margreta, 99, 182 n. 3
*De interpretatione* (Aristotle), 55–56
*De inventione dialectica* (Agricola), 63
*De magistro* (Augustine), 20–21, 92
de Man, Paul, 102, 108, 170
*De ratione studii* (Erasmus), 62–63, 92
Derrida, Jacques, 170
*De sapientia veterum* (Bacon): fables in, 16, 22–23, 79–81, 84, 88, 95–99, 171, 183 n. 28; on Homer and Aristotle, 22–23; political allegory in, 95–99, 106, 107, 110–12; reform of knowledge in, 78, 80
Discourse theory, 108–9, 169, 184 nn. 31, 32. *See also under* Ricoeur
Dubois, Paige, 15, 26

Else, G. F., 32
Elsky, Martin, 99–101, 108, 110, 179 n. 3, 183 n. 25
Epistemes (of knowledge), 53–54, 99, 101, 117, 168, 174 n. 12. *See also under* Foucault; Reiss
Erasmus, Desiderius: and Bacon, 89, 91–92; and Boyle, 133; on St. John's Gospel, 61–62; *sermo*, 21, 61–63 (*see also under* Language Theory); on sign systems, 19. Works of: *De copia*, 63–65; *De ratione studii*, 62–63, 92

*Essay Towards a Real Character and a Philosophical Language, An* (Wilkins), 130–31
*Essays* (Bacon), 77, 181 n. 18; "Of Simulation and Dissimulation," 73–74; "Of Truth," 76
*Experientia literata*: in *Advancement of Learning*, 81–85, 100, 170; in *De dignitate et augmentis*, 81–85, 96, 104; in *Instauratio magna*, 171; in *Redargutio philosophiarum*, 89–90

Fables: in *Advancement of Learning*, 80, 88; in *De sapientia veterum*, 16, 22–23, 79–81, 84, 88, 95–99, 171, 183 n. 28; in Hobbes' *Leviathan*, 122, 128; political use of, 101, 107, 110–12; in scientific *mimesis*, 91–92; in *Temporis partus masculus*, 78, 80. *See also under* Bacon; Hobbes.
Farrington, Benjamin, 77
Ferguson, Margaret, 70
Fish, Stanley, 19–20, 142–43
Foucault, Michel: on epistemic knowledge, 53–54, 99, 168, 174 n. 12

Gilman, Ernest B., 138, 152
Glanville, Joseph: on moral certainty in science, 17 (*see also under* Science); on witch craft, 47
Godzich, Wlad, 171
Goldberg, Jonathan, 102
Gombrich, E. H., 29, 190; *Art and Illusion*, 29, 48

Haraway, Donna, 25
Hardison, O. B., 179 n. 39
Hartlib Circle, 134–36, 184 n. 3. *See also under* Boyle; Jacobs, J. R.
Havelock, Eric, 34–35
Heraclitus, 34
Hieroglyphics, 82–83, 96, 108, 131. *See also under* Bacon
*History of the Royal Society* (Sprat), 115–17, 185 n. 7
Hobbes, Thomas, 13, 117–30; allegory, use of, 117–19; dispute with Boyle, 17, 117, 123, 132, 133–34; on diversity of speech, 124–29 (*see also under* Speech); on fables, 122, 128; on humans as artificers, 118–19, 126 (*see also under* Machines); on imagination and memory, 119–21, 125 (*see also* Imagination; Memory); on language theory, 31–32, 117–18, 123, 125, 127; *Leviathan*, 31, 117–30; and Milton, 141, 147–48, 162, 188 n. 17
Hoopes, Robert, 188 nn. 26 and 30
Horkheimer, Max, 14–15
Hovey, Kenneth Alan, 179 n. 41, 183 n. 18
Howell, A. C., 31
Howell, W. S., 66
Hulse, Clark, 107, 183 nn. 23 and 26
Humanist pedagogy, 52–53
Hunter, J. Paul, 186 n. 19

Imagination, 23, 114, 119–21, 125. *See also* Hobbes, imagination and memory; *Leviathan*, imagination; Memory; Science, imagination in; Sprat, imagination in science
Imitation Theory, 32–36, 68–70, 73, 138, 181 n. 22; *apate* (trickery), 33–34, 36, 138, 155–65, 181 n. 23 (*see also under* Mimesis). *See also under* Pre-Socratics; Sidney
*Instauratio magna* (Bacon), 22, 76; *experiential literata* in, 171 (see also under *Advancement of Learning, De dignitate et augmentis, Redargutio philosophiarum*); induction, 98 (*see also under* Reason); influences of, 130
Irigaray, Luce: *Speculum of the Other Woman*, 156–57, 160–61, 164–65
Iser, Wolfgang, 168

Jacob, J. R.: Hartlib Circle and the Invisible College, 134–36, 184 n. 3
Jansenists, 139
Jardine, Lisa, 109, 180 n. 13
Johnston, David, 185 n. 12
Jones, Robert Foster, 184 n. 2

Keller, Evelyn Fox, 26
Kneale, William and Martha, 177 n. 9

Lacan, Jacques, 153; mirror stage, 157–58
Language: logic in language, 61; occult languages, 23 (*see also under* John

Language (*continued*)
  Webster); philosophical grammar, 104, 182 n. 3 (*see also* Universal languages); on scientific metaphor, 130–32;
Language theory, 21, 31–32, 59–60, 61–63, 117–18, 123, 125, 127, 133–34. *See also* Boyle, language theory; Erasmus, *sermo*; Hobbes, language theory; Peter of Spain, *significatio*, *suppositio*
Lanham, Richard, 19–20
*Leviathan* (Hobbes), 31, 117–30; allegory, use of, 117–19; imagination in, 119–21, 125; language theory, 117–18, 123, 125, 127; man as machine, 118–19, 126; memory in, 119–21, 125; political allegory in, 118, 122, 128; on speech, 124, 126–27, 128, 129
Lyons, John D., 138, 152, 154

Machine (man as), 25, 118–19, 126. *See also* Haraway; Hobbes, humans as artificers; *Leviathan*, man as machine
Martin, Julian, 85–86, 94, 99, 180 n. 14
Memory, 119–21, 125. *See also* Hobbes, imagination and memory; Imagination; *Leviathan*, memory
Merchant, Carolyn, 15, 26, 173 n. 3
Methods of knowledge: exoteric and acroamatic methods, 87–88; magistral and probative methods, 84–86. *See also under* Bacon
Milton, John, 13, 139–65; and Bacon, 25, 144–45; on discursive reason, 139–40, 148 (*see also under* Reason); and Hobbes, 141, 147–48, 162, 188 n. 17; on intuitive reason, 24–25, 149–50, 162 (*see also under* Reason); *mimesis* in, 146–47, 155–65; on Ramism, 140, 142; on *res et verba*, 25, 143–44. Works of: *Areopagitica*, 150; *Art of Logic*, 140–41; *Paradise Lost*, 24–25, 139–40, 145–65; *Paradise Regained*, 147–48, 151, 154–55; *Reason of Church Government Urged Against Prelaty*, 141–45, 149
*Mimesis* (Resemblance), 68, 114; *apate* (trickery), 33–34, 36, 138, 155–65; 181 n. 23; in Aristotle, 39–41; in Bacon, 49, 90–91, 101; dramatic *mimesis*, 19, 33, 35, 50; in Milton, 146–47, 155–65; in Plato, 32–39, 46; in the pre-Socratics, 32, 35; in Quintilian, 181 n. 23; and *res et verba*, 20, 31–39; in Ricoeur, 39–41, 90–91, 175 n. 20. *See also under* Bacon; Milton; Plato; Pre-Socratics, imitation theory; Quintilian; *Res et verba*; Ricoeur
Myth. *See* Fables

Norbrook David, 183 n. 23
*Novum organon* (Bacon), 131, 169; experimental method in, 88–89; reform in, 76, 78; utilitarianism in, 19

Ong, Walter, 178 nn. 18, 23 and 24; on Ramism, 140–41, 187–88 n. 13

Padley, G. A., 182 n. 3, 185 n. 11; on universal languages, 129–32, 185 n. 16
Parable. *See* Fables
*Paradise Lost* (Milton), 145–65; feminism in 164–65; intuitive and discursive reason in, 24–25, 139–40, 148, 149–50, 162 (*see also under* Milton; Reason); resemblance, 145–47, 157–60, 163. *See also under* Anamorphism
*Paradise Regained* (Milton), 147–48, 151, 154–55. *See also under* Anamorphism
Pascal, Blaise, 139, 152
Patterson, Annabel, 101
Pérez-Ramos, Antonio, 16, 173 nn. 5, 7, 8 and 9; on the ethics in science, 113 (*see also under* Science); on the maker's knowledge tradition, 18 (*see also under* Science)
Peter of Spain: and Aquinas, 60; *Summulae logicales*, 56, 58–61; on *suppositio* and *significatio*, 59–60 (*see also under* Language theory)
Pico della Mirandola, Giovanni, 177 n. 34
Plain style, 23, 24, 114, 137. *See also under* Royal Society

Plato: and *mimesis*, 32–39, 46; *Republic*, 35, 37–38, 46; and Sidney, 67–68
*Poesis*, 30, 68–69, 72, 112, 174 n. 13. *See also* Sidney, philosophy, *poesis*
Pre-Socratics: on allegory, 104–5; Bacon on, 78–80, 96, 104–5, 169; on imitation theory, 32, 35
*Prior & Posterior Analytics* (Aristotle), 56
Puttenham, George, 21

Quilligan, Maureen, 102
Quint, David, 177 n. 34
Quintilian: and Bacon, 31, 91; and Hobbes, 117; on *mimesis* 181 n. 23

Rabinow, Paul, 25–26, 171–72
Ramism, 140–42; 187–88 n. 13. *See also* under *Art of Logic*; Milton; Ong; Sidney
Reason: analogical reasoning, 23, 131–32 (*see also under* Boyle); discursive reason, 139–40, 148 (*see also under* Milton); induction, 98 (see also under *Instauratio magna*); intuitive reason, 24–25, 149–50, 162 (*see also under* Milton)
*Reason of Church Government Urged Against Prelaty* (Milton), 141–45; poet as truth-giver, 143–44; reader-response theory, 142–43; right reason, 141–42, 145, 149
*Redargutio philosophiarum* (Bacon), 78; *experientia literata* in, 89–90 (See also under *Advancement of Learning, De dignitate et augmentis, Instauratio magna*)
Reiss, Timothy: on epistemic knowledge, 53–54, 101, 117, 174 n. 12
Relational semantics, 48–49, 99, 117, 176 n. 31, 187 n. 5. *See also under* Waswo
*Republic* (Plato), 35, 37–38, 46
*Res et verba* (words and things): 13, 62, 65, 107, 139; in Bacon, 19, 97, 168, 170; in humanist pedagogy, 53; in Milton, 25, 143–44; and *mimesis*, 20, 31–39; in the Royal Society, 114, 168; and structuralism, 44–46, 48
*Richard II* (Shakespeare), 107, 151–52
Ricoeur, Paul, 43; on Aristotle, 39–40, 175 n. 19; on discourse theory, 108–9, 169, 184 nn. 31 and 32; on *mimesis*, 39–41, 90–91, 175 n. 20; *Rule of Metaphor*, 39–44
Rosenmeyer, Thomas G., 35–36
Rossi, Paolo, 80–81; on historiography of science, 54, 99 (*see also under* Science)
Royal Society: on civic humanism, 24, 47; influential members, 17, 113; on philosophical grammar, 182 n. 3; on plain style, 23, 24, 114, 137; on *res et verba*, 114, 168
*Rule of Metaphor* (Ricoeur), 39–44; on *mimesis*, 40–41; on scientific models, 39, 43–44
*Rule of Reason* (Wilson), 65–67

Salmon, Vivian, 185–86 n. 16
Sargent, Rose-Mary, 134, 186 n. 21
de Saussure, Ferdinand, 44; *Course in General Linguistics*, 44, 176 n. 24
Schmitt, Charles B., 52–53, 177 n. 3
Scholasticism, 61, 66. *See also under* Valla; Wilson
Schwartz, Regina, 151
Science: ethics in, 113 (*see also under* Pérez-Ramos); historiography of, 54, 99 (*see also under* Baconian Science; Rossi); imagination in, 23, 114 (*see also under* Sprat); maker's knowledge and, 18 (*see also under* Pérez-Ramos); metaphor in, 46–47, 116–17, 126 (*see also under* Sprat); moral certainty in, 17 (*see also under* Glanville); on scientific demonstration, 58–60 (*see also under* Aquinas).
Shakespeare, William, 107; *Richard II*, 107, 151–52
Shapiro, Barbara, 20, 174 n. 11
Shuger, Debora, 168
Sidney, Sir Philip: *Defence of Poetry*, 51, 67–73; on imitation theory, 68–70, 73, 181 n. 22; on philosophy, 68–69, 72; and Plato, 67–68; on *poesis*, 30, 68–69, 112, 174 n. 13; and Ramism, 141
Sign systems, 19–21, 92–94, 169. *See also under* Augustine; Erasmus
Slaughter, M. M., 137
Sloan, Thomas, 140

*Some Considerations about Reason and Religion* (Boyle), 134, 186 n. 18
*Speculum of the Other Woman* (Irigaray): and the female reader, 164–65; on Descartes, 156–57, 161; on narcissism, 160–61
Speech, 21, 45–46, 61–63, 105–6, 124–29. *See also* Erasmus, *sermo*; Hobbes, diversity of speech; Speech-act theory
Speech-act theory, 45–46, 105–6. *See also under* Cicero; Speech
Spenser, Edmund: and Bacon, 74–75, 107, 112; on civic humanism 136; *Faerie Queene*, 74
Sprat, Thomas: *History of the Royal Society*, 115–17, 185 n. 7; on imagination in science, 23, 114 (*see also under* Imagination; Science); on metaphor in science, 46–47, 116–17, 126 (*see also under* Science); as monarchist, 17, 47–48
Stephens, James, 183 n. 25
Structuralism, 44–46, 48. See also under *Res et verba*
*Summulae logicales* (Peter of Spain): reform of logic and dialectic, 56, 61; *scientia sermocinalas*, 58–59; *supppositio* and *significatio*, 59–60 (*see also under* Language Theory)

*Temporis partus masculus* (Bacon), 78, 80
*Things Said to Transcend Reason* (Boyle), 134

*Topics* (Aristotle), 55
Tourangeau, Roger, 42
Tuve, Rosemond, 175–76 n. 21

Universal languages, 50, 104, 129, 130–32, 182 n. 3, 185 n. 16. See also *An Essay Towards a Real Character*; Language, on scientific metaphor, philosophical grammar; Padley, on universal languages

Valla, Lorenzo: on scholasticism, 61
Van Etten, Henry, 153
Vickers, Brian, 115, 121
Vives, Juan Luis, 61

Warhaft, Sidney, 99, 181 n. 3
Waswo, Richard: relational semantics, 48–49, 99, 117, 176 n. 31, 187 n. 5
Webster, Charles, 185 n. 16
Webster, John, 17; on occult languages, 23
Whitman, Jon, 106
Whitney, Charles, 180 n. 13
Wilkins, John, 130–32; and Bacon, 130–31; *An Essay Towards a Real Character and a Philosophical Language*, 130–31
Wilson, Thomas: and Bacon, 66; on rhetoric, 65; *The Arte of Rhetorique*, 65–67; *Rule of Reason*, 65–67; on the political use of rhetoric, 21; on scholasticism, 66